T0214347

SpringerBriefs in Applied Sciences and Technology

PoliMI SpringerBriefs

More information about this series at http://www.springer.com/series/11159
http://www.polimi.it

Riccardo Casati

Aluminum Matrix Composites Reinforced with Alumina Nanoparticles

POLITECNICO
DI MILANO

Springer

Riccardo Casati
Department of Mechanical Engineering
Politecnico di Milano
Milan
Italy

ISSN 2191-530X   ISSN 2191-5318   (electronic)
SpringerBriefs in Applied Sciences and Technology
ISSN 2282-2577   ISSN 2282-2585   (electronic)
PoliMI SpringerBriefs
ISBN 978-3-319-27731-8   ISBN 978-3-319-27732-5   (eBook)
DOI 10.1007/978-3-319-27732-5

Library of Congress Control Number: 2015958311

Printed on acid-free paper

This Springer imprint is published by SpringerNature
The registered company is Springer International Publishing AG Switzerland

Preface

A lot of efforts have been made by the scientific community to create new materials with optimal combinations of properties. The capability to design and create specific microstructures is a fundamental requirement to make materials able to show the desired functional and mechanical properties. However, the design of the right combination of physical, mechanical, or functional characteristic of materials has turned out to be difficult to achieve, because these properties may barely coexist in engineering materials. For example, it is desirable to create light materials that combine both high strength and good toughness, although strength and toughness are generally mutually exclusive in metals. Hence, only some combination of properties are possible, and compromises are often needed.

In December 1959, the famed lecture *"There's Plenty of Room at the Bottom,"* which was presented at Caltech by the Nobel Prize winner Richard Feynman, gave birth to a new branch of science: Nanotechnology. From that event on, the interest in nanoscale structures has grown rapidly among scientists and within the industrial community. Nanoengineering and nanoscience have led to an explosion of new experimental techniques for assembling and manipulating nanostructures. The availability of nano-sized particles as a result of one of these new developments, new sintering processes, and characterization techniques have brought about the creation of different feasible scenarios for preparing new materials with extremely interesting mixtures of properties. From this perspective, nanocomposite materials are believed to be potential candidates for many applications that require low weight, high mechanical strength, good ductility and toughness, damping capability, and microstructural stability at high temperature. Unfortunately, the production of such materials is not trivial with conventional melting and casting techniques. Powder metallurgy, on the other hand, provides flexibility in terms of materials selection and microstructure design. Plastic deformation-based techniques, such as hot extrusion and equal channel angular pressing (ECAP), can achieve consolidation at relatively low temperatures and shorter times than conventional sintering processes; then they are able to maintain any nanostructure created in the particles. A fine microstructure is generally considered a desirable aspect for metals: ultrafine-grained and nanocrystalline materials revealed indeed unprecedented high strength, often several times higher than that of their coarse-grained counterparts. With this work, several efforts have been made to better understand the role of the processing parameters to produced ductile light metal nanocomposites with improved mechanical and damping properties. In particular, Al matrix composites with ultrafine microstructure reinforced with either in-situ or ex-situ alumina nanoparticles were produced through a combination of powder processing by ball milling and powder consolidation by either ECAP or hot extrusion.

Contents

Summary and Aim of the Work

Aluminum alloys show remarkable properties, such as low density, good resistance to corrosion, and low thermal expansion. These characteristics make them attractive materials for several industrial fields where important applicative constrains have to be satisfied. For example, lightweight (higher performance and lower consumption) and improved mechanical and functional properties (strength, corrosion, and wear resistance) are essential features that materials have to possess in order to be employed in many applications in the mechanical, automotive, and aerospace field. Another important feature of Al alloys is their recyclability, as reprocessing does not damage their structure. Moreover, CO_2 emission limitations and energy cost make lightweight materials a priority condition.

In this view, Al-based metal matrix composites are considered very interesting. These hybrid materials show capabilities to design lightweight structures with a precise balance of mechanical and physical properties, with a relevant improvement on the tribological characteristics, and also high temperature strength. Furthermore, the reinforcement particles are generally thermodynamically stable at the elevated temperatures, making these materials suitable for high temperature applications. A novel concept of composites, which further enhances the properties of conventional composites, is given by the design of metals reinforced by nanoparticles. Due to their very small size, the nano-fillers are able to interact with the lattice defects, i.e., dislocations, enabling new strengthening mechanisms to be activated. Their impact can be of great relevance from either the scientific or the technological point of view. As metal matrix nanocomposites (MMnCs) are a very novel class of materials, the lack of knowledge associated with them is still to be filled up. Several technological issues have to be overcome in order to produce bulk nanocomposites characterized by homogeneous dispersion of nanoparticles and high mechanical performance. The comprehension of the physical phenomena related to their improved mechanical behavior and functional properties is still incomplete and needs deeper understanding.

The aim of this work consisted in the development of Al nanocomposites with enhanced damping and mechanical properties and good workability. The nanocomposites exhibited high strength, good ductility, improved damping behavior, and the capability of being worked into wires. Since the production of MMnCs by conventional melting processes was considered to be extremely critical because of the poor wettability of the nanoparticles, different alternative powder metallurgy routes were adopted. Alumina nanoparticles were embedded into Al powders by severe grinding and consolidated using several techniques. Special attention was directed to the structural characterization at micro and nanoscale as uniform nanoparticles dispersion in metal matrix is primarily important. Moreover, some of the billets produced via powder metallurgy were also rolled to prepare wires as an example of the final product. The Al nanocomposites revealed an ultrafine microstructure reinforced with alumina nanoparticles produced in-situ or added ex-situ.

The work had a strong empirical basis. Different sintering methods and parameters were employed to produce MMnCs characterized by well-dispersed nanoparticles in the Al matrix. In particular, different powder metallurgy routes were investigated, including high energy ball milling and unconventional compaction methods (ECAP, BP-ECAP, hot extrusion). The physical, mechanical, and functional behavior of the produced materials was then evaluated by different mechanical tests (hardness tests, instrumented indentation, compressive and tensile tests, dynamo-mechanical analysis) and microstructure investigation techniques (scanning and transmission electron microscopy, electron back scattering diffraction, X-ray diffraction, differential scanning calorimetry). The experimental results were then theoretically discussed. Literature equations and models were also used to predict the mechanical behavior of the material and the numerical and experimental results were compared.

Chapter 1
State of the Art of Metal Matrix Nanocomposites

Abstract In this chapter, the state of art of metal matrix nanocomposites (MMnCs) is discussed. In particular, the sintering methods used so far by researchers to prepare different combinations of metal matrix and nanoparticles as well as the possible applications of nanocomposites are described. The strengthening mechanisms involved in the extraordinary mechanical properties of MMnCs are also introduced, reporting the literature formulas used to calculate their contributions to the final strength of the material. In this work, Al-based composites reinforced with nanoparticles were prepared via powder metallurgy methods, first by grinding different mixtures of powders via high-energy ball milling, then by compacting powders through equal channel angular pressing (ECAP) and hot extrusion. Therefore, these processing techniques are thoroughly described in this introductory study.

Keywords Metal matrix nanocomposites · Strengthening mechanisms · Reinforcement · Ball milling · Powder metallurgy · ECAP · Extrusion

1.1 Metal Matrix Nanocomposites

Metal matrix nanocomposites are a new class of materials consisting of two or more physically and/or chemically distinct phases [1, 2]. The composites generally own some superior characteristics than those of each of the individual components. A number of processing routes are available for the synthesis of nano-reinforced metal matrix composites. They are based either on solid sintering or on liquid processing. Consolidation of powder, generally preceded by high-energy ball milling, can be carried out both by conventional technique (hot isostatic pressing, forging or cold isostatic pressing followed by heat treatment) or by alternative methods, such as ECAP or hot extrusion. Among the liquid processes, promising results were achieved by ultrasonic assisted casting. MMnCs are very interesting materials with high potential for use in a large number of industrial applications. Some recent research works highlighted the real possibility to produce composites

© The Author(s) 2016   1
R. Casati, *Aluminum Matrix Composites Reinforced with Alumina Nanoparticles*,
PoliMI SpringerBriefs, DOI 10.1007/978-3-319-27732-5_1

characterized by exciting mechanical properties, which can be further enhanced by optimizing the particle dispersion. In particular, remarkable results in terms of hardness, mechanical strength, wear resistance, creep behavior and damping properties were achieved in several pioneering research works [1, 2]. By the adoption of this class of composites, expensive heat treatment (typically solution annealing and precipitation aging) currently carried out on conventional monolithic alloys could be avoided and the range of available alloys for structural and functional applications could be broadened. Notwithstanding their potential properties, there are still some aspects to be improved in production of metal matrix composites (MMCs) reinforced with nanoparticles. Their fabrication is much more complicated than that of conventional composites reinforced with fibers or micro-reinforcements. When the particles scale down from the micro- to the nano-level, many additional difficulties have to be solved and new issues have to be faced. The reaction between ceramic nanoparticles or carbon nanotubes with the matrix is still unclear. The inappropriate bonding interface may lead to the failure of the composites. Clustering of particles is another issue of paramount importance to be solved, especially for the production of large parts [1, 2].

1.1.1 Type of Metal Matrices and Reinforcements

Several metallic materials have been considered as matrix constituent for the preparation of MMnCs. In particular, the most interesting metals for industrial applications are Al [3–28], Mg [29–37], Ti [38–40], Cu [41–44] and their alloys. Pure and alloyed aluminum is among the most investigated materials with the largest number of published research studies describing Al-based composites as possible candidates for structural applications. Different species of nano-sized oxides (Al_2O_3, Y_2O_3) [8, 15, 21–23, 43, 45], nitrides (Si_3N_4, AlN) [34], carbides (TiC, SiC) [4, 13, 16, 18, 24, 25, 28, 30–33], hydrides (TiH_2) [36] and borides (TiB_2) [17, 42] have been employed as reinforcement agents. Moreover, different allotropes of carbon (carbon black [7], fullerenes [37] and carbon nanotubes (CNTs) [3, 19, 20, 35, 41, 44, 46]) have been investigated as fillers for several research works published in literature. The most used particles are CNTs: if well dispersed they are able to confer very high mechanical properties to the metal matrix and can lead to increased electrical conductivity, which makes MMnCs very attractive materials for electrical and electronic applications. Single wall carbon nanotubes (SWCNT) and multi-wall carbon nanotubes (MWCNT) have been both used for MMnCs production. In this regard, for example copper-0.1 wt% MWCNT composites revealed a 47 % increase in hardness and bronze-0.1 wt% SWCNT showed a 20 % improved electrical conductivity [41]. Finally, intermetallic compounds (NiAl, Al3Ti) had also been successfully used as reinforcement phase in MMnCs [14, 47]. Al-Al3Ti nanocomposite revealed good mechanical behavior at high temperature [47], while TiAl-NiAl MMnCs showed low fracture toughness and very high hardness [14].

1.1.2   Strengthening Mechanisms

The high strength of metal matrix nanocomposites is the result of several strengthening mechanism contributions, namely:

 i. load-transfer effect (or load-bearing effect), which is due to the transfer of load from the metal matrix to the hard reinforcement [4];
 ii. Hall-Petch strengthening (or grain boundaries strengthening), which is related to the grain size of the metal matrix. Nanoparticles may play a fundamental role in matrix grain refinement [48–50];
 iii. coefficient of thermal expansion (CTE) and elastic modulus (EM) mismatch, which are responsible of creating dislocation networks around the particles [51, 52].
 iv. Orowan strengthening, due to the capability of nanoparticles to obstacle the dislocation movement [50, 53, 54];

Moreover, like any other metallic material, the MMnCs can be further strengthen by Hull and Bacon [50]:

 v. work hardening (or strain hardening or cold working), i.e. plastic deformation of metal, which leads to dislocation multiplication and development of dislocation substructures;
 vi. solid-solution hardening, which can be obtained by adding interstitial or substitutional atoms in the crystal lattice which are responsible for the deformation of the lattice itself and for the formation of internal stresses;
 vii. precipitation hardening (or age hardening), which relies on changes in solid solubility with temperature, to produce fine precipitates which impede the movement of dislocations, or defects in a crystal lattice. Dislocations can cross the particles by cutting them or they can bow around them by the Orowan mechanism.

Thus, several concurrent effects contribute to the final strength of MMnCs. The strengthening effects are superimposed and connected to each other; therefore, it is not yet clear how to attribute to the single effects the right contribution (weight) to the final strength. Nevertheless, several methods to predict the final strength of the nanocomposites were proposed in the open literature.

In the following sections, the strengthening methods correlated to the addition of nanoparticles in the metal matrix (i.e. points of the list from i to iv), as well as the models for estimating the final strength of the nanocomposites are described.

1.1.2.1   Load Transfer Effect

The transfer of load from the soft and compliant matrix to the stiff and hard particles under an applied external load, contributes to the strengthening of the base material. A modified Shear Lag model proposed by Nardone and Prewo [4] is commonly used to predict the contribution in strengthening due to load transfer in particulate-reinforced composites [52–54]:

$$\Delta\sigma_{LT} = v_p\sigma_m \left[\frac{(l+t)A}{4l} \right] \tag{1.1}$$

where v_p is the volume fraction of the particles, σ_m is the yield strength of the unreinforced matrix, l and t are the size of the particulate parallel and perpendicular to the loading direction, respectively. For the case of equiaxed particles [52], Eq. (1.1) reduces to:

$$\Delta\sigma_{LT} = \frac{1}{2}v_p\sigma_m. \tag{1.2}$$

1.1.2.2 Hall-Petch Strengthening

The grain size has a strong influence on metal strength since the grain boundaries (GBs) can hinder the dislocation movement. This is due to the different crystal orientation of adjacent grains and to the high lattice disorder characteristic of these regions, which prevent the dislocations from moving in a continuous slip plane. Impeding dislocation movement, GBs hinder the extensive onset of plasticity and hence increase the yield strength of the material. When an external load generates a shear stress in a material, existing dislocations and new dislocations move across the crystalline lattice until facing a GB, which creates a repulsive stress field to oppose the dislocation movement. Then, dislocation pile up occurs generating extensive repulsive stress fields that act as a driving force to reduce the energetic barrier for their diffusion through the boundary. A decrease in grain size leads to a decrease in the amount of extensive pile ups (Fig. 1.1). Then the necessary load to be applied for dislocation movement through the material must be higher. The higher the applied stress needed to move the dislocations, the higher the yield strength [48, 49].

The Hall-Petch equation relates the strength with the average grain size (d) [38, 48, 49]:

$$\Delta\sigma_{H-P} = \frac{k_y}{\sqrt{d}} \tag{1.3}$$

where k_y is the strengthening coefficient (characteristic constant of each material). The particles play a fundamental role in final grain size found in metal matrices of composites since they can interact with grain boundaries acting as pinning points, retarding or stopping their growth on high-temperature processing. The increase of v_p (volume fraction) and the decrease of d_p (particle diameter) lead to a finer structure, as theoretically modeled by the Zener equation [52]:

Fig. 1.1 Schematic of the Hall-Petch strengthening mechanism. Dislocations are sketched using the reversed symbol "T"

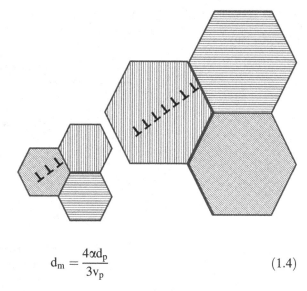

$$d_m = \frac{4\alpha d_p}{3v_p} \tag{1.4}$$

where α is a proportional constant.

There is a limit to this strengthening mechanism. For grain sizes lower than a threshold value d^* [55–66], the size of dislocations begins to approach the size of the grains, prohibiting dislocation wide-ranging pile-ups and instead resulting in grain boundary sliding or rotating, resulting in a decrease in the material's yield strength and an in increase in ductility (superplasticity). Therefore, the mechanical behavior of nanocrystalline materials is said to deviate from the classical Hall-Petch relation, below which the k value gradually decreases as the grain size is reduced (i.e. less effective strengthening). Eventually, nanocrystalline materials may soften, as the grain size is further reduced, and this effect is termed the inverse Hall-Petch relationship (often observed in metals having a crystallite size of a few units of nanometer).

1.1.2.3 Orowan Strengthening

The so-called Orowan mechanism consists in the direct interaction of nano-particles with dislocations. The non-shearable ceramic particles pin the crossing dislocations and promote dislocation bowing around the particles (Orowan loops) under external load (Fig. 1.2) [50].

The Orowan effect can be evaluated by the following expression:

$$\Delta\sigma_{OR} = \frac{0.13bG}{d_p\left(\sqrt[3]{\frac{1}{2}v_p} - 1\right)} \ln\left(\frac{d_p}{2b}\right) \tag{1.5}$$

where b is the Burger's vector and G is the matrix shear modulus.

Fig. 1.2 Schematic of the
Orowan strengthening
mechanism [50]

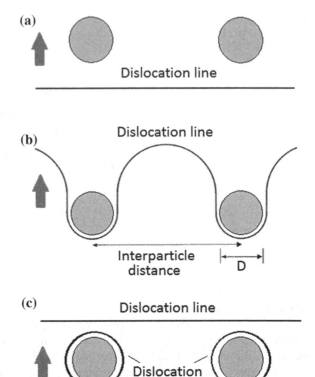

Orowan strengthening is more relevant to MMnCs with particle sizes smaller than 100 nm [52–54, 67, 68]. This is because larger sized particles lead to large interparticle distance for the same volume fraction of particles and tend to segregate to the GBs. Under these circumstances, the contribution of Orowan bowing mechanism becomes negligible.

1.1.2.4   CTE and EM Mismatch

Even after perfectly matching processing conditions (amount of given deformation, times and temperatures of processing and heat treating) much higher dislocation density exists in the matrix of a composite than in the unreinforced matrix due to thermal residual stresses. The increase in dislocation density from the contribution of residual plastic strain develop during post processing cooling as a result of different coefficient of thermal expansion (CTE) between the matrix and reinforcing phase. The high stress field around the reinforcement is relaxed by the generation of dislocations at the matrix-reinforcement interface. Such dislocations are also called geometrically necessary dislocations (GNDs). Also the gap in elastic modulus

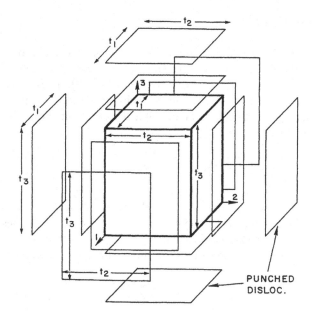

Fig. 1.3 Schematic of the model for "punching" of prismatic dislocation loops to relax the CTE misfit at the matrix/particulate interface [70]

(E) between the stiff particles and the compliant matrix can lead to the formation of additional dislocations during the elastic straining of the nanocomposites. GND density due to CTE (ρ^{CTE}) and Young modulus (ρ^{EM}) mismatch can be estimated by the following expressions [51, 52, 69, 70]:

$$\rho^{CTE} = \frac{A\Delta\alpha\Delta T v_p}{b d_p (1 - v_p)} \tag{1.6}$$

$$\rho^{EM} = \frac{6 v_p}{\pi d_p^3} \varepsilon \tag{1.7}$$

where A is a geometric constant, $\Delta\alpha$ is the difference in CTE and ΔT is the difference between test and processing or heat treatment temperatures. Arsenault and Shi [70] firstly proposed Eq. 1.6, which was supported by the so called "dislocation punching model" (Fig. 1.3). The combined strengthening due to CTE and E GNDs can be calculated by means of the Taylor equation [71]:

$$\Delta\sigma_{CTE+EM} = \sqrt{3}\beta G b \left(\sqrt{\rho^{CTE}} + \sqrt{\rho^{EM}} \right) \tag{1.8}$$

where β is a constant.

1.1.2.5 Sum of Contributions

The final strength of the MMnC is quite difficult to be estimated; to this purpose, different models have been proposed. The easiest one does not take into account the overlapping of the concurrent strengthening effects [52]. It evaluates the final strength of the composite, σ_c, by simply summing the contributions related to the single strengthening effects, $\Delta\sigma_i$, with the original yield strength of the unreinforced matrix, σ_m, therefore:

$$\sigma_c = \sigma_m + \sum_i \Delta\sigma_i \qquad (1.9)$$

Other studies proposed alternative methods to calculate σ_c, considering the superposition of the effects. A simple model [51, 52], which approaches quite well the experimental data, suggests to calculate the final strength of the composite by summing the root of the squares of all the single strengthening contributions, as:

$$\sigma_c = \sigma_m + \sqrt{\sum_i \Delta\sigma_i^2} \qquad (1.10)$$

Another common method that takes into account Orowan strengthening effect, dislocation density due to the residual plastic strain caused by the CTE mismatch and load-bearing effect was proposed by Zhang and Chen [53, 54]:

$$\sigma_c = (1 + 0.5v_p)(\sigma_m + A + B + \frac{AB}{\sigma_m}) \qquad (1.11)$$

where A is the term relative to CTE mismatch and B is the coefficient related to Orowan effect:

$$A = 1.25G_m b\sqrt{\frac{12\Delta\alpha\Delta Tv_p}{bd_p(1 - v_p)}} \qquad (1.12)$$

$$B = \frac{0.13G_m b}{d_p\left[\left(\frac{1}{2v_p}\right)^{\frac{1}{3}} - 1\right]} \ln\frac{d_p}{2b} \qquad (1.13)$$

Few papers are available in literature about this topic. This lack does not allow a comprehensive evaluation and comparison of the proposed methods based on a physics ground.

1.1.3   Preparation Methods and Properties

For the large-scale production of metal matrix nanocomposites, the main problem to face is the low wettability of ceramic nano-particles, which does not allow the preparation of MMnCs by conventional casting and melting processes since the result would be an inhomogeneous distribution of particles within the matrix. The high surface energy (related to very high surface-to-volume ratio) readily leads to the formation of clusters of nanoparticles, which are not effective in hindering the movement of dislocations and can hardly generate a physical-chemical bond to the matrix, thus reducing significantly the strengthening capability of nanoparticles [1, 46]. Several unconventional production methods have been studied by researchers in order to overcome the wettability issue, either by formation of the reinforcement by in situ reactions or by ex situ addition of the ceramic reinforcement by specific techniques. Hereafter, the most studied and successful methods are described by classifying them into liquid, semisolid and solid processes.

1.1.3.1   Liquid Processes

For composites prepared by the conventional liquid metallurgy route, severe aggregation of nanoparticles frequently occurs even when mechanical stirring is applied on the melt before casting. This is due to poor wettability and high viscosity generated in the molten metal owing to high surface-to-volume ratio of the nano-sized ceramic particles. The density of nanoparticles do not play an important role in the production process of nanocomposites. Such small particles are supposed to float on the top of the molten bath even if their density is relatively higher than that of the liquid matrix. This mass-mismatch issue was indeed very significant in micron-sized particle reinforced composites but in nano-reinforced materials, other effects such as those induced by extensive surface tension play a much more important role [72].

High-intensity ultrasonic waves revealed to be useful in this context since they produce acoustic transient cavitation effects, which lead to collapsing of micro-bubbles. The transient cavitation would thus produce an implosive impact, strong enough to break the nanoparticle clusters and to uniformly disperse them in the liquid metal. According to this technique, a good dispersion of 2 % of SiC nano-particles in aluminum alloy 356 was achieved by Li et al. [13] by means of the experimental setup equipped by ultrasonic source. An improvement of 20 % in hardness over the unreinforced alloy was achieved. Lan et al. produced nano-sized SiC/AZ91D Mg alloy composites through the same method (see the experimental device in Fig. 1.4). A fairly good dispersion of the particles was achieved although some small clusters still occurred into the matrix. Owing to general improvement of the dispersion, the 5 wt% SiC reinforced composite led to a microhardness increase of 75 % [73]. As already mentioned, the nanoparticles also play a fundamental role in grain refinement, working as pinning points, hampering the grain growth and leading to improved mechanical properties according to Eqs. (1.3) and (1.4).

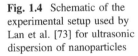

Fig. 1.4 Schematic of the experimental setup used by Lan et al. [73] for ultrasonic dispersion of nanoparticles

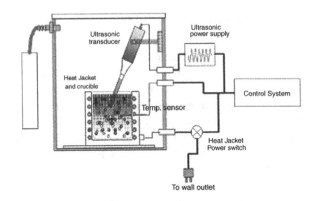

In this regard, it has been reported that an addition of 1 wt% nano-SiC into pure Mg strongly acts in this direction. Under comparative processing conditions, the Mg/SiC composite featured an average grain size of 72 μm whereas the unreinforced pure Mg showed an average size of 181 μm [30]. Moreover, De Cicco and co-workers [27, 28] proved by a droplet emulsion technique (DET) that nanoparticles can catalyze nucleation, thereby reducing undercooling. For A356 alloy based nano-composites produced by ultrasonic assisted casting, γ-Al$_2$O$_3$ revealed a better nucleation catalyzer than α-Al$_2$O$_3$ probably due to its lower lattice mismatch with the metal matrix. Other tests were also conducted in the same research [28] with TiC and SiC of different sizes.

Tensile tests performed on AZ91D alloy and on the same material reinforced by 1 wt% of nano-AlN produced by ultrasound-assisted casting revealed an increase of yield strength in MMnCs of 44 % at room temperature (RT) and of 21 % at 200 °C when compared to the unreinforced AZ91D alloy. For the same materials, a decrease of fracture strain at room temperature (RT) was achieved while an enhanced ductility was measured at 200 °C [34]. Improved ductility was also detected by Wang et al. [31] even at RT. The yield strength (YS), ultimate tensile strength (UTS) and fracture elongation of an AZ91 alloy were 104, 174 MPa and 3.6 %, respectively whereas the corresponding values for the AZ91 alloy reinforced by 0.5 wt% of 50 nm SiC were: 124, 216 MPa and 6.6 %, respectively. In a research work by Cao et al. [74], the addition of 1.5 wt% SiC to Mg-4Zn alloy obtained by an ultrasonic cavitation-based solidification process led to an increase of RT ductility of more than twice as well as to improved YS and UTS. The same authors also observed a reduction of grain size in reinforced sample (150 μm vs. 60 μm) which was also related to improved castability of the alloy.

Disintegrated melt deposition (DMD) is a further liquid metallurgy process successfully employed for nano-composite production (Fig. 1.5). Alumina nanoparticles have been well dispersed in Al–Mg alloys by heating the metal in argon atmosphere and adding the ceramic particles by means of a vibratory feeder. The melt was stirred and poured, then disintegrated with argon gas jets and deposited onto a metallic substrate. Finally, the MMnCs were extruded to reduce porosity down to very low levels and to achieve a good dispersion of the particles [75, 76].

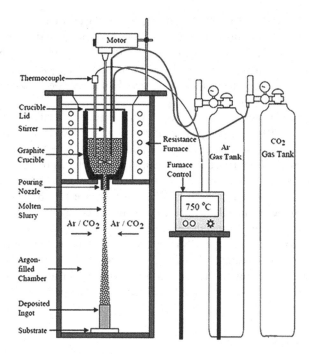

Fig. 1.5 Schematic of the experimental setup used for disintegrated melt deposition (DMD) process [75]

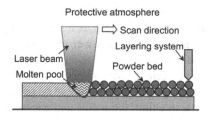

Fig. 1.6 Schematic of selective laser melting process [77]

Selective laser melting (Fig. 1.6) was also used to produce Ti-based composites reinforced by nanoparticles [77]. Powders were milled by high-energy ball milling and then melted by laser beam under protective atmosphere. Through this method, a unique material featuring a very different microstructure of the reinforcement was achieved. A proper decrease in volumetric energy density led to the development of TiC whiskers and of uniformly dispersed nano-lamellar TiC starting from dendritic TiC. The same research confirmed that well dispersed nano-particles induce improved mechanical and wear properties to the Ti matrix.

MMnCs reinforced with nanoparticle formed in situ have been successfully prepared by liquid metallurgy processes. 50 nm TiB_2-reinforced copper-matrix composites were produced by adding B_2O_3, C and Ti in a Cu-Ti melt [78]. The composites exhibited significantly improved mechanical properties. In particular, the YS of Cu and Cu/TiB$_2$ was 298.7 and 509.6 MPa, respectively. Al/TiB$_2$ nanocomposites were also synthesized by an in situ method, by adding a mixture of potassium hexafluorotitanate (K_2TiF_6) and potassium tetrafluoroborate (KBF_4) salts in an Al melt under argon atmosphere [79].

High-pressure die casting [35] and arc-discharge plasma method [80] were also used to produce AZ91/CNT composites and in situ Al/AlN MMnCs, respectively.

Finally, it was highlighted that the main problem to be faced in production of CNT-MMnCs by the liquid metallurgy method is the interaction of the nanotubes with the liquid metal. In fact, the process may cause damage to CNTs or formation of chemical reaction products at the CNT/metal interface [46, 81, 82]. Therefore, this synthesis route is mainly indicated for those composite matrices having low-melting temperatures and reduced reactivity with the reinforcement phases. The problem of low wettability of CNTs can be partially overcome by coating CNT with metal layers (for example Ni) [46, 83]. The field of surface modification appears as quite promising and it is open to innovation for attenuating the drawbacks about wettability and tendency to clustering of nanoparticles.

1.1.3.2   Semi-Solid Processes

Only few works are available in literature about this topic even if this method has been widely applied for micrometer-size particle-reinforced MMCs, and it would be extremely interesting for large-scale production.

A356/Al$_2O_3$ MMnCs were produced by using a combination of rheocasting and squeeze casting techniques [84]. Rheocasting is a semi-solid phase process, which has several advantages: it is performed at lower temperatures than those conventionally employed in foundry practice resulting in reduced thermochemical degradation of the reinforcement surface. Moreover, the material shows thixotropic behavior typical of stir cast alloys and production can be performed by conventional foundry methods. During rheocasting, the pre-heated nanoparticles are added in the semi-solid slurry while it is vigorously agitated in order to achieve a homogenous particle distribution. Then the slurry is squeezed using a hydraulic press. Mg alloy AZ91 ingots reinforced by nano-SiC particles were produced by semisolid stirring-assisted ultrasonic vibration [33]. After homogenization treatment and extrusion, the SiC reinforcement featured a fairly good dispersion although bands of accumulated nanoparticles were present and their amount could be reduced by increasing the extrusion temperature.

An innovative method named semi-solid casting (SSC) was proposed by De Cicco et al. (Fig. 1.7) [85]. Zinc alloy AC43A reinforced by 30 nm β-SiC was used for sample preparation by SSC. The SSC experiments were carried out by pouring ultrasonicated molten MMNC material (450 °C) from a graphite crucible into a steel injection device, which was preheated to 400 °C. Liquid MMnC was cooled down to 386 °C achieving less than 30 % of solid fraction. Then, the injection sleeve was

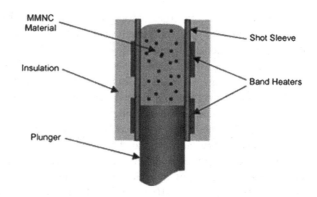

Fig. 1.7 Schematic of experimental setup of the semi-solid casting (SSC), which was proposed by De Cicco et al. in [85]

MMNC Material

Insulation

Plunger

Shot Sleeve

Band Heaters

inverted and placed on top of a steel mold. The plunger was activated and the semi-solid material was injected into the mold. The produced samples showed strength properties comparable to those produced by ultrasound-assisted casting but with improved ductility due to the nucleation catalysis by the β-SiC nanoparticles that refines the microstructure of the MMnC.

1.1.3.3   Solid Processes (Powder Metallurgy)

Several solid methods were studied and developed for preparing MMnCs. In particular, different powder metallurgy techniques were successfully employed in this respect. Some papers focus on mechanical alloying which is a powder metallurgy technique consisting in repeated cold welding, fracturing and re-welding of powder particles in a high-energy ball mill. This technique is of fundamental importance since it allows achieving a better dispersion of nano-powder into the composite by breaking up the ceramic clusters. It can also be exploited for the formation of alloys by diffusion mechanisms starting from pure metals, and to produce preforms by in situ reaction of nano-reinforcements. Therefore, mechanical alloying, which cannot be separated from the opportunity of breaking up the nano-ceramic clusters, is a value-added option offered by this particular processing route [22, 24, 28, 36, 42, 86–95]. It has been proved that the presence of nanoparticles can accelerate the milling process (stimulating plastic deformation, cold welding, and particle fragmentation) and grain refinement mechanism [89–91]. Process control agent (PCA) has a strong influence on morphological evolution of powders during ball milling [23]. The addition of 1.5 % stearic acid as PCA prevents extensive cold welding of Al particles during ball milling and leads to an increase of hardness of the hot-compacted samples. Speed and time of milling, mass of balls and powder, ball diameter also contribute to final hardness development. In particular, a pronounced decrease in energy transfer from the balls to the powder was found by raising the amount of balls [24].

High-energy ball milling proved to be a suitable technique for the production of in situ MMnCs. Al-TiN composite was prepared by milling elemental Al and Ti powders with ring-type organic compound pyrazine in benzene solution [87]. Mg 5 wt% Al alloy in situ reinforced with TiH_2 was also prepared by mechanical alloying of elemental powder of Mg, Al and Ti, using polyethylene-glycol to provide hydrogen for the formation of TiH_2 and to prevent excessive cold welding during ball milling. After attritioning, the powders were cold isostatically pressed (CIP), extruded and thermal treated. The mechano-chemically milled specimens showed very fine microstructure and good dispersion of fine reinforcements, a slight increase in YS and ductility was observed [36, 92]. Iron-wustite (Fe–FeO) nanocomposites were also produced by mechano-chemical processing starting from Fe and Fe_2O_3 powder with different mole ratios. These materials showed a ferromagnetic-like behavior, which was interpreted according to spinel-like defect, clusters [88]. Mg-5%Al-10.3%Ti-4.7%B (wt.) powders were ground using high-energy ball milling and extruded by Lu et al. They observed the formation of non-equilibrium Ti_3B_4 phase in extruded samples [90]. Lu and co-authors investigated the in situ formation of TiB_2 via chemical reaction among Al, TiO_2 and B_2O_3. The powders were cold compacted into green compacts and sintered at different temperatures. By this method, 53 % improvement in both YS and UTS was achieved [17]. In situ TiB_2 reinforced Cu alloy composite was indeed achieved via argon atomization at 1400 °C followed by hot isostatic pressing (HIP) at 200 °C under 200 MPa pressure [42]. Moreover, $Cu-Al_2O_3$ nanocomposites have been prepared by two chemical routes: through decomposition of $Al(NO_3)_3$ to Al_2O_3 by calcination of a paste of $CuO-Al(NO_3)_3$ followed by H_2 reduction and sintering, or through hydrolysis of $Al(NO_3)_3$ solution followed by calcination, reduction and sintering. The latter method led to the formation of finer Al_2O_3 (30 nm vs. 50 nm) nanoparticles and promoted enhanced properties in terms of relative density, microhardness and abrasive wear resistance [86]. Submicron-sized titanium carbide was successfully sintered from the reaction of Ti salt (K_2TiF_6) and activated carbon, by controlling the degree of reaction through temperature and amount of C. In this respect, it was observed that at low temperatures, formation of Al_3Ti was predominant while at high temperatures (above 1000 °C), the intermetallic compound was not stable and TiC was preferentially formed [25].

Several techniques have been used to perform the compaction of composite powders. The most common routes are HIP [42], hot pressing [43] and cold pressing [15, 17, 45, 93–95] or CIP [36, 92] followed by a sintering treatment. Conventional hot extrusion [19, 32, 96, 97] or equal channel angular extrusion (ECAE), also known as equal channel angular pressing (ECAP) [7, 8, 10, 21, 23, 26, 44, 98–100] revealed to be suitable methods to achieve full dense composites. Hot extrusion was used to sinter Al-2 % CNT composite powders blended by high-energy ball milling, observing a tensile strength enhancement of 21 %. Extrusion was also found to promote alignment of CNTs along the extrusion direction that may lead to anisotropic properties of the material [19]. In the same work, CNTs have been found to act as nucleation sites for void formation during tensile tests. Both CNTs pullout and MW-CNT inner tubes slippage were observed in fractured surfaces, suggesting poor interfacial bond between CNTs and Al matrix. Ferkel et al. extruded at 350 °C pure Mg powders and two composite powders consisting of Mg and 3 wt% nano-SiC. One batch had been ball milled and the second one had been conventionally mixed [32]. The study was focused on high temperature mechanical behavior of the produced nanocomposites. The milled composite showed the largest gain in strength but also the lowest ductility at all testing temperatures (RT, 100, 200 and 300 °C). Moreover, significant difference in the creep response was been observed at 200 °C in favor of the ball-milled composite. Al-based samples sintered by ECAP and sintered by cold pressing followed by heat treatment and extrusion were compared in [21]. The best results were achieved by the former method since the hardness values after three ECAP passes was 67 % higher than the extruded samples. Higher compressive strength and increased wear resistance were also achieved in the ECAP processed samples. ECAP powder pressing was successfully used to consolidate 1 % CNTs in copper matrix at room temperature, avoiding CNT surface reaction with metal matrix [44]. Al nanocomposites reinforced by carbon black (CB) or by Al_2O_3 were also produced by using back-pressure ECAP at 400 °C [7, 8, 10]. In particular, good dispersion of Al_2O_3 and CB nanoparticles was achieved by mechanical milling followed by 8 ECAP passes. Compression tests performed on these materials showed that the YS of unreinforced sample reached 58 MPa, while that of the composite with addiction of 5 % CB reached 260 MPa. Moreover, after 8 ECAP passes, fully dense pure Al showed a Vickers hardness of 37.1 HV, while the Al-5 % Al_2O_3 MMnC showed an hardness of 96.5 HV and the Al-5 % CB system an hardness of 81 HV [7, 10]. Al composites reinforced by 5, 10 and 15 % nano-Al_2O_3 were also produced by powder metallurgy route. The powders were mixed in ethanol by ultrasonic treatment, wet attritioned by high-energy ball milling and finally compacted by ECAP at 200 °C. The best results in terms of microhardness and compressive yield stress were achieved by adding 10 % nano-alumina after 4 ECAP passes [26]. Milling associated to ECAP process of chips revealed to be a further promising route for the use of metal scraps [38, 99]. Ceramic nanoparticles (AlN) were successfully added to Mg-5 % Al alloy chips even though no significant improvement in strength could be achieved [98].

Since in this work MMnCs have been produced via powder metallurgy route, first by grinding the powders by high-energy ball milling, then by compacting them via ECAP or hot extrusion, in the next sections these processes will be thoroughly debated.

1.2   Processing of Metal Powders

There are several methods employed for processing and consolidating metal powders. Most of the consolidation methods are based on the sintering process, which relies on diffusion of atoms between powder particles at fairly high temperatures. Other methods are indeed based on plastic deformation of metal powders. Hereafter, high-energy ball milling, which is a technique for grinding and modifying powders, and ECAP and extrusion, which are process used for consolidation of powders, are described in details being of particular interest for this work.

1.2.1   High-Energy Ball-Milling

Mechanical alloying is a powder processing technique that allows the production of homogeneous materials starting from blended elemental powder mixtures. It was firstly developed by John Benjamin in 1966 [101]. A popular technique to perform mechanical alloying is the high-energy ball milling (BM). It consists in vials filled with powders and grinding balls, which can rotate with planet-like motion at very high speed. BM produces nanostructured particles through structural decomposition/refinement of initially coarse grained structures by severe plastic deformation (SPD) [101, 102]. Minimum grain sizes measuring few tens of nanometers can be easily obtained in ball milled powder particles. Moreover, BM is a versatile powder processing technique that is capable of producing powder with unique and far-off equilibrium microstructures. During high-energy milling, the powder particles are repeatedly attritioned, cold welded, fractured and rewelded. The force of the impact plastically deforms the powder particles trapped within two colliding balls, leading to extensive work hardening and fracture. The new fresh surfaces enable the particles to weld together leading to an increase in particle size. In a second stage of the process, the particles get work hardened and fatigue fractures may occur. At this stage, the tendency to fracture predominates over cold welding. After milling for a certain time, equilibrium is established between the welding and the fracturing rates. Smaller particles are able to withstand deformation without fracturing and tend to be welded into larger clusters, with an overall tendency to drive both very small and very large particles towards an intermediate size [101].

It is clear that during mechanical alloying, severe deformation is introduced into the powder particles. This is readily proven by the high amount of crystal defects, such as dislocations, vacancies, stacking faults, and grain boundaries. The presence of this defects also increase the diffusivity of solute elements into the matrix. Moreover, the rise in temperature during milling, which can be controlled by milling parameters, further helps the diffusion processes, and consequently promotes effective alloying amongst the constituent elements [101, 103, 104]. It is possible to conduct mechanical alloying of three different combinations of materials:

- Ductile-ductile system. The condition for alloying to take place during BM is fracturing and cold-welding of the mixture to introduce crystal defects such as dislocations, vacancies, stacking faults, and grain boundaries. This leads to enhanced diffusion of solute elements into the matrix and decreased diffusion distances. Thus, the ductile-ductile system is the easiest combination to achieve by mechanical alloying since cold welding cannot take place if the particles are too brittle [101, 103, 105].

Fig. 1.8 Schematics
depicting **a** the ball motion
inside the mill and **b** the
collision ball-powder-ball
during mechanical alloying
[101]

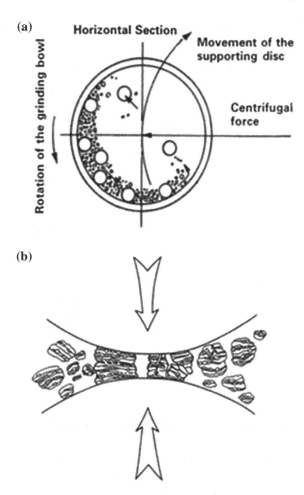

- Ductile-brittle system. Ceramic and intermetallic particles are generally brittle.
  Fracture into smaller fragments occurs during early stages of BM [101, 106, 107],
  as illustrated in Fig. 1.8. The fragmented brittle particles tend to be enveloped by
  the ductile particles and are trapped at the inter-lamellar interfaces, as shown in
  Fig. 1.9a. Further milling convolutes and refines the ductile particles (Fig. 1.9b
  and c). The interlamellar spacing decreases and the brittle particles are dispersed
  more homogeneously. Alloying between the ductile and brittle phase may occur
  depending on the solid solubility of the system.
- Brittle-brittle systems. Alloying is less likely to occur during BM of
  brittle-brittle powder mixtures due to lack of welding. On the contrary, BM has
  a strong influence in powder particles size owing to marked fragmentation
  effects [101].

Fig. 1.9 Structural evolution
of ductile-brittle system
during BM for oxide
dispersion strengthening
alloys [101]

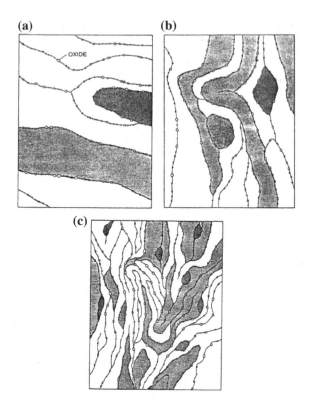

1.2.1.1   High-Energy Ball Milling Variables

A number of variables must be optimized to achieve the desired product phase
and/or microstructure. Hereunder the most important are listed:

- Material of the vials. The material used for the milling container is important
  since due to impact of the grinding balls on the inner walls of the vial, some
  material will be dislodged and incorporated into the powder. This can con-
  taminate the powder. Hardened C-steel, tool steel, hardened chromium steel,
  tempered steel, stainless steel, WC-Co, are the most common types of materials
  used for the grinding vessels. Some specific materials are used for specialized
  purposes; these include copper, titanium, sintered corundum, yttria-stabilized
  zirconia (YSZ), partially stabilized zirconia-yttria, agate, etc. [108–114].
- Shape of the vials. The shape of the container is also important, especially the
  internal design of the container, but cylindrical vials are generally employed.
- Milling speed and direction. The faster the rotation the higher is the kinetic
  energy transferred to the powder, but above a critical speed the balls are pinned
  to the vial inner walls and do not fall down to apply any impact force. Therefore,
  the maximum speed should be just below this threshold value so as to produce
  the maximum collision energy. At high speed, the temperature of the

vial-ball-powder system may reach a high value, which can be advantageous or disadvantageous depending on the purpose of the milling. Same or opposite direction of the main disk and of the planets can be chosen, this may lead to different collision configurations [101].

- Milling time and pauses. Milling time has a fundamental role on the characteristic of the final powder. The effect of the milling time on the average particle size was described in the previous part of the section. In order to decrease the temperature of the system, a certain number of pauses can be programmed. Milling times can become extremely long (several tenths of hours) to achieve the desired transformation of the powder.
- Material of the grinding balls. The materials used for the balls are the same of those used for the vials, because they possess the same criticality.
- Size of the grinding balls. Balls size has an influence on the milling efficiency. A large size of the grinding medium is useful since the larger weight will transfer more energy to the particles. It was also suggested that smaller balls produce intense frictional action, which promotes the amorphous phase formation. In fact, it appears that soft milling conditions (small ball sizes, lower energies, and lower ball-to-powder ratios) seem to favor the amorphization of the crystal structure or the formation of metastable phases [108, 115–118]. Also, different sizes balls can be used simultaneously. It has been reported that a combination of large and small size balls during milling minimizes the amount of cold welding and the amount of powder stuck onto the surface of the balls [119].
- Ball-to-powder weight ratio (BPR). The ratio of the weight of the balls to the powder is an important variable in the milling process. The higher the BPR, the shorter is the milling time required. At a high BPR, because of an increase in the weight proportion of the balls, the number of collisions per unit time increases and consequently more energy is transferred to the powder particles. It is also possible that due to the higher energy, more heat is generated and this could also change the constitution of the powder.
- Extent of filling the vial. It is necessary that there is enough space for the balls and the powder particles to move around freely in the container in order to have enough impact energy.
- Milling atmosphere. To avoid contamination, either argon or helium are generally used as protective gas. Nitrogen, air and hydrogen can lead to the in situ formation of nitride, oxide and hydride compounds.
- Process control agent. The severe plastic deformation of powder particles during milling leads to their cold-welding, but a good balance between cold welding and fracturing of particles is necessary to achieve alloying and/or good particles dispersion. A process control agent (PCA) (also referred to as lubricant or surfactant) is often added to the powder mixture to reduce the effect of welding and to inhibit the agglomeration. PCAs can be in the solid, liquid, or gaseous form. They are mostly organic compounds, which are able to act as surface-active agents. The lubricant adsorbed on particle surfaces obstructs the cold welding and lowers the surface tension of the powders. Several PCAs have

been employed at a content level of about 1–5 % of the total powder weight. The most used PCAs include stearic acid, hexane, methanol, and ethanol. Most of these lubricants decompose and form compounds with the powders during milling, which are incorporated as inclusions or dispersoids into the powder particles during milling. These are not necessarily harmful to the alloy system since they can contribute to dispersion strengthening of the material, resulting in increased strength and higher hardness. The hydrogen subsequently escapes as a gas or it is absorbed into the metal lattice on heating or sintering. Even though hydrogen gas primarily serves as a surfactant and does not usually participate in the alloying process, some reports indicate that hydrogen acts as a catalyst for amorphous phase formation in titanium-rich alloys [120–123].

- Temperature of milling. The temperature of milling can be varied by dripping liquid nitrogen on the milling container or by electrically heating the milling vial. It was reported that higher temperature results in lower strain in the material and hence bigger grain size [124, 125]. Moreover, milling temperature can modify the amorphization kinetics [126, 127].

1.2.2 Equal Channel Angular Pressing

ECAP is a widely researched SPD process able to induce large accumulative shear strain on a bulk or powdered material without significantly modifying the specimen geometry. The ECAP was firstly proposed by Segal in the former Soviet Union in the early 1970s [128]. During 1990s, with growing interest in ultrafine grained materials, ECAP was recognized as a capable SPD method and the research on ECAP gained momentum [129–131]. One of the main advantages of ECAP over other SPD processes is that it allows relatively large billets to be processed with homogeneous deformation throughout the sample.

Grain refinement by ECAP has been widely studied and many comprehensive reviews on the properties and structural evolution of ECAP processed materials are available [132–139]. ECAP was also used to refine the grains of MMC materials [140, 141]. The presence of the particles accelerates the hardening and grain refinement of the matrix phase compared to the unreinforced material. This was attributed to their role in retaining higher dislocation densities at the particle interface and in stimulating continuous recrystallization processes [141].

The basic setup for ECAP is depicted in Fig. 1.10 [133]. It consists of a die having two intersecting channels with equal cross section and a plunger that fits the entrance channel of the die. Two angles are characteristic of the ECAP die:

- The angle Φ, made by the intersect of the two channels
- The angle ψ, which describe the outer corner curvature where the two channels intersect.

Fig. 1.10 Schematic drawing
of an ECAP setup [133]

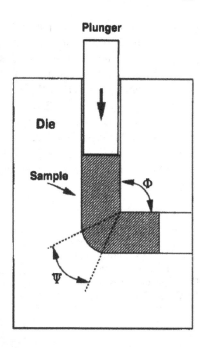

A billet, with the same cross section as the channel, is forced by a plunger to
flow through the die. It is deformed by simple shear as it passes through the plastic
deformation zone (PDZ) located at the intersection of the two channels.

For a round corner die, i.e. when $\psi \neq 0$, the equivalent shear strain at each pass
can be calculated using the following equation [134]:

$$\varepsilon_e = \frac{1}{\sqrt{3}} \left[2 \cot\left(\frac{\Phi}{2} + \frac{\psi}{2}\right) + \psi \mathrm{cosec}\left(\frac{\Phi}{2} + \frac{\psi}{2}\right) \right] \tag{1.14}$$

It is worth noting that the Eq. (1.14) is only valid under ideal conditions (i.e. simple
shear). The cumulative plastic strain (or effective strain) after N number of passes
can be expressed as

$$\varepsilon_{eN} = \varepsilon_{eN} * N \tag{1.15}$$

It can be easily estimated that remarkably high values of accumulated strain (of the
order of several units) can be achieved by repeating more times the insertion of a
billet in the ECAP die, so as to attain the expected grain refining effects.

1.2.2.1   ECAP process variables

The final characteristic of the material produced by ECAP can be tailored modifying the following process parameters:

- Geometry of the die. In particular the intersect angle Φ, and the outer angle ψ affect the equivalent strain, as previously described.
- Processing route. Different microstructures can be attained or final refinement can be achieved at different rates by simply rotating the workpiece between subsequent passes [136, 142]. Four basic processing routes are commonly employed. They are defined according to their rotating scheme around the vertical axis, namely: route A, route B_A, route B_C, and route C, as illustrated in Fig. 1.11a [142]. The different routes are able to impose different shear planes on the sample (Fig. 1.11b) [136]. It has been demonstrated that route B_C leads to the fastest transformation from subgrain boundaries to an equiaxed array of high

Fig. 1.11 a The four common processing routes used in ECAP [142]. **b** Shear planes for consecutive passes of the four fundamental ECAP routes viewed on the X, Y, and Z planes [136]

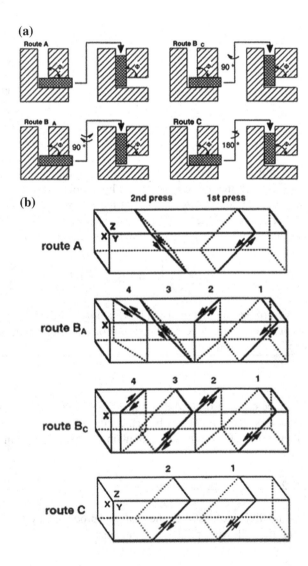

angle grain boundaries (HAGBs) by promoting the development of subgrain bands through subsequent ECAP passes on the intersecting sets of the two different shear planes [137].

- Temperature. In general, the higher the working temperature, the bigger the final grain size of the sample and the lower the achieved yield strength. Samples processed at elevated temperatures also tend to have higher fractions of low angle grain boundaries (LAGBs). During ECAP deformation, LAGBs evolve into HAGBs by absorption of moving dislocations into the boundaries. At elevated temperatures, the rate of recovery is higher and dislocations are more likely to annihilate by cross-slip rather than being absorbed into subgrain boundaries. Therefore, formation of HAGBs is reduced when the working temperature is high [143–146].
- Number of passes through the die. A minimum amount of plastic deformation (i.e.: a minimum number of passes) is required to achieve the expected microstructural evolution leading to significant grain refinement. Usually, a condition of saturation of properties is attained after a given amount of deformation which depends on nature of material, crystal structure, processing temperature and route [136, 137, 142].
- Back-pressure (BP). A constant BP can be applied on the workpiece by the use of a back-plunger [7, 147, 148]. This leads the material to be subjected to a hydrostatic pressure. BP plays only a minor role in final mechanical and microstructure properties [149]. However, it modifies the mode of deformation during ECAP closer to that of simple shear by eliminating the formation of a corner gap, improving the uniformity of the stress-strain distribution and reducing the tendency to form cracks in brittle materials [5, 150]. Applying BP implies the formation of overall more equiaxed microstructure with higher fraction of HAGBs (about +10 %) [151]. The benefits of BP are most evident when processing brittle materials which normally require high processing temperatures [152–155]: they can be processed at much lower temperatures, resulting in much finer grain structures.

1.2.2.2 Consolidation of Particles by ECAP

ECAP was first used for powder consolidation in the 1990s [156]. This technique was used to consolidate different materials starting from powders such as Al, Ti, Cu, and Mg, alloyed powders and composite powders. Traditional sintering methods are based on diffusion. Vice versa, the mechanism for ECAP consolidation is mainly by plastic deformation of the powder particles [99]. During ECAP, the particles undergo very high shear strains and plastic deformation. The high deformation breaks the brittle surface oxide films and leads to direct contact between fresh metal surfaces, and bonding between the particles can occur instantaneously (Fig. 1.12) [99].

Fig. 1.12 Schematic
illustration of particle
consolidation in conventional
sintering (*left*) and ECAP
consolidation (*right*) [99]

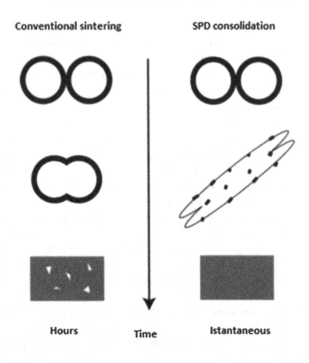

Fig. 1.13 Tensile curves for
pure bulk (IM) and powdered
(PM) Al samples
processed/consolidated by
ECAP [147]

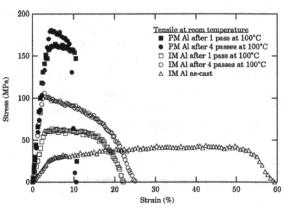

Fully dense bulk material are produced thanks to the material flow accompanying
the deformation, which is believed to be able to close the pores between the particles
[157]. Since ECAP is not a high-temperature slow-diffusion consolidation process, it
can be performed at much lower temperatures and shorter processing times than
conventional sintering process. ECAP consolidated samples showed remarkably
high yield strength and hardness compared to their ingot counterparts due to their
fine and equiaxed microstructure [6, 147, 158], as shown in Fig. 1.13 [147].

The application of BP is very useful during ECAP consolidation of powdered materials at low temperatures, especially when cans are not used. Shearing of individual particles and fracturing of the surface oxide layer are the prerequisite for SPD powder consolidation [6]. With the application of a BP, the particles are pushed against each other, generating more friction when passing the shear zone and causing the particles to shear deform more efficiently instead of just sliding on each other.

1.2.3 Hot Extrusion

Extrusion is a process used to produce parts with fixed cross-section (bars, wires, billets, profiles, tubes). There are two main types of extrusion processes:

- direct extrusion: a plunger pushes a workpiece forward through a die, causing a reduction in cross-sectional area of the workpiece.
- indirect or inverted extrusion: the metal is contained in a blind cylinder, the plunger which houses the die compresses the metal against the container, forcing it to flow backward to the die in the hollow plunger.

The extrusion process can be done at low or, more frequently, at high temperatures to improve material plasticity. The great advantage of the extrusion process lies in the relative ease and efficiency with which complex sections can be produced in one operation. A schematic representation of the direct extrusion process is depicted in Fig. 1.14. The reduction of cross section (reduction ratio, R) of the extruded part can be calculated as follow:

$$R = \frac{A_0}{A_f} \tag{1.16}$$

where A_0 and A_f are the cross areas of the billet before and after extrusion, respectively [159].

Fig. 1.14 Schematic of the direct extrusion process

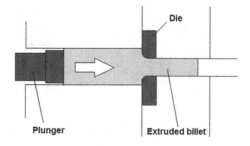

1.2.3.1 Extrusion Process Variables

The force needed for extrusion is affected by a variety of circumstances. Mainly, it is affected by the nature of the material, by the temperature and by the speed at which the extrusion is performed. Moreover, the amount of deformation involved must be considered, i.e. the reduction ratio. Additional factors are the shape of the section and the design of the die used. For pure aluminum and the soft alloys, for example, reduction ratios of up to 100:1 are not uncommon, while for the hard alloys the ratio is usually between 8:1 and 40:1. The speed of extrusion depends upon the available pressure and thus on the alloy temperature and reduction ratio [159].

1.2.3.2 Consolidation of Particles by Extrusion

Pioneering experiments on the extrusion of metal powders were carried out in the late 1950s [160]. In principle, both direct and indirect extrusion methods may be used to extrude metal powders, but direct extrusion is much more widely used. For this reason, this section will only focus on direct extrusion. During extrusion, metal powders undergo plastic deformation, usually at high temperatures, to produce a full dense form. The three main routes to extrude powders are depicted in Fig. 1.15. In the first method, a hot container supplies heat to the charge and extrusion is performed without atmosphere protection. The second method relies on the use of a precompacted billet, referred as a "green compact", as the extrusion workpiece.

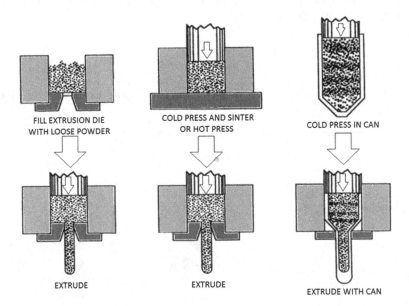

Fig. 1.15 Three main routes to extrude metal powders [161]

Cold or hot pressing are suitable process for producing green compacts. However, sintering is not a required processing step and many materials are extruded without this additional processing. The third approach consists in partially densifying the metal powder in a can. This may then be evacuated and sealed, or it may be left open to the atmosphere [161].

Canning can be employed for the following advantages [162]:

- isolation of the powdered material from the atmosphere and lubricants: it is a clean extrusion technique;
- isolation of toxic materials for safe handling;
- encapsulation of spherical and other difficult-to-compact powders to produce a billet form;
- reduction of friction and facilitation of metal flow at the die interface by proper selection of the can material;
- isolation of the powdered material from the extrusion channel, which is the region of highest shear. It is an important consideration for materials with limited ductility.
- capability of positioning powders and solid components within the can to produce unique and complex shapes.

When purity must be maintained, canning of the powder, including gas evacuation and sealing, is an essential step [97, 163]. Procedures may include glove box container preparation and assembly, total isolation of powder from ambient air, evacuation at slightly elevated temperatures to drive off adsorbed gases from particle surfaces, and leak checking of sealed containers. Sometime, a requirement is that the extruded product must be free from both particle native oxide layers and nonmetallic inclusions that can degrade some mechanical properties, like fracture toughness and fatigue resistance [161, 162].

1.3 Possible Applications of MMnCs

So far, to the author' best knowledge, metal matrix composites reinforced by nanoparticles or nanotubes are not being employed yet in relevant commercial applications due to their very recent development. However, MMnCs show higher mechanical properties than micro-particles reinforced composites, without any evidence of a strong drop in thermal and electrical conductivity [5, 6]. For this reason, they are considered as possible candidates for substituting conventional MMCs or related monolithic alloys in structural and electrical applications at room and high temperatures. For example, CNT composites could replace, thanks to their higher strength and stiffness, carbon fibers composite in many applications, especially in high-temperature environments. Another good opportunity for the substitution of traditional MMCs with nano-sized counterparts is related to the good fracture toughness and ductility, which was the main Achille's heel in micro-reinforced MMCs. Toughness can be substantially preserved in nano-reinforced composites owing to the reduced particle volume fraction required to achieve strengthening, thus widening the opportunity for structural applications, even in highly demanding and safety parts.

For instance, the enhanced wear resistance [9] and the good thermal conductivity combined to the high specific strength make MMnCs attractive materials for aircraft brakes. Moreover, the specific strength and elastic modulus could be exploited in sport industry, for instance for rackets or bicycle frames and other components. A further field of potential application is in electronic devices, for example for heat sinks and solders (thanks to their thermal properties) or as antennas (thanks to their electrical properties and stiffness). Aerospace and automotive industries may exploit all the above properties for different kind of applications such as structural radiators, gears, aircraft fins, cylinder liners, pistons, disk brakes and calipers.

The improved damping capacity of MMnCs could also be exploited to reduce vibrations and noise of structures. In Mg-Al_2O_3 samples extruded at 350 °C after powder milling, a significant damping ability was highlighted and attributed to interface character of MMnCs [7].

References

1. S.C. Tjong, Novel nanoparticle-reinforced metal matrix composites with enhanced mechanical properties. Adv. Eng. Mater. **9**, 639–652 (2007)
2. R. Casati, M. Vedani, Metal matrix composites reinforced by nano-particles—a review. Metals **4**(1), 65–83 (2014)
3. C.F. Deng, D.Z. Wang, X.X. Zhang, Y.X. Ma, Damping characteristics of carbon nanotube reinforced aluminum composite. Mater. Lett. **61**, 3229–3231 (2007)
4. V.C. Nardone, K.M. Prewo, On the strength of discontinuous silicon carbide reinforced aluminum composites. Scr. Metall. **20**, 43–48 (1986)
5. C. Xu, K. Xia, T.G. Langdon, The role of back pressure in the processing of pure aluminum by equal-channel angular pressing. Acta Mater. **55**, 2351–2360 (2007)
6. X. Wu, K. Xia, Back pressure equal channel angular consolidation—Application in producing aluminum matrix composites with fine flyash particles. J. Mater. Process. Technol. **192–193**, 355–359 (2007)
7. S. Goussous, W. Xu, X. Wu, K. Xia, Al-C nanocomposites consolidated by back pressure equal channel angular pressing. Comput. Sci. Technol. **69**, 1997–2001 (2009)
8. W. Xu, X. Wu, T. Honma, S.P. Ringer, K. Xia, Nanostructured Al–Al_2O_3 composite formed in situ during consolidation of ultrafine Al particles by back pressure equal channel angular pressing. Acta Mater. **57**, 4321–4330 (2009)
9. Y. Li, Y.H. Zhao, V. Ortalan, W. Liu, Z.H. Zhang, R.G. Vogt, N.D. Browning, E.J. Lavernia, J.M. Schoenung, Investigation of aluminum-based nanocomposites with ultra-high strength. Mater. Sci. Eng. A **527**, 305–316 (2009)
10. S. Goussous, W. Xu, K. Xia, Developing aluminum nanocomposites via severe plastic deformation. J. Phys: Conf. Ser. **240**, 012106 (2010)
11. M. Kubota, X. Wu, W. Xu, K. Xia, Mechanical properties of bulk aluminium consolidated from mechanically milled particles by back pressure equal channel angular pressing. Mater. Sci. Eng. A **527**, 6533–6536 (2010)
12. F. He, Q. Han, M.J. Jackson, Nanoparticulate reinforced metal matrix nanocomposites—a review. Int. J. Nanopart. **1**, 301–309 (2008)

13. X. Li, Y. Yang, X. Cheng, Ultrasonic-assisted fabrication of metal matrix nanocomposites. J. Mater. Sci. **39**, 3211–3212 (2004)
14. S.X. Mao, N.A. McMinn, N.Q. Wu, Processing and mechanical behavior of TiAl/NiAl intermetallic composites produced by cryogenic emchnical alloying. Mater. Sci. Eng. A **363**, 275–289 (2003)
15. H. Mahboob, S. A. Sajjadi, S.M. Zebarjad, Syntesis of Al-Al$_2$O$_3$ nanocomposite by mechanical alloying and evaluation of the effect of ball milling time on the microstructure and mechanical properties, in *Proceedings of International Conference on MEMS and Nanotechnology (ICMN '08)*. Kuala Lumpur, Malaysia, pp. 240–245, 13–15 May 2008
16. M. Gupta, M.O. Lai, C.Y. Soo, Effect of type of processing on the microstructural features and mechanical properties of Al-Cu/SiC metal matrix composites. Mater. Sci. Eng. A **210**, 114–122 (1996)
17. L. Lu, M.O. Lai, Y. Su, H.L. Teo, C.F. Feng, In situ TiB$_2$ reinforced Al alloy composites. Scripta Mater. **45**, 1017–1023 (2001)
18. M. Gupta, M.O. Lai, M.S. Boon, N.S. Herng, Regarding the SiC particulates size associated microstructural characteristics on the aging behavior of Al-4.5 Cu metallic matrix. Mater. Res. Bull. **33**, 199–209 (1998)
19. A.M.K. Esawi, K. Morsi, A. Sayed, A. Abdel Gawad, P. Borah, Fabrication and properties of dispersed carbon nanotube-aluminum composites. Mater. Sci. Eng. A. **508**, 167–173 (2009)
20. C.F. Deng, D.Z. Wang, X.X. Zhang, A.B. Li, Processing and properties of carbon nanotubes reinforced aluminum composites. Mater. Sci. Eng. A. **444**, 138–145 (2007)
21. R. Derakhshandeh Haghighi, S.A. Jenabali Jahromi, A. Moresedgh, M. Tabandeh Khorshid, A comparison between ECAP and conventional extrusion for consolidation of aluminum metal matrix composite. J. Mater. Eng. Perform. **21**, 1885–1892 (2012)
22. C. Carreño-Gallardo, I. Estrada-Guel, M. Romero-Romo, R. Cruz-García, C. López-Meléndez, R. Martínez-Sánchez, Characterization of Al$_2$O$_3$ NP–Al2024 and AgCNP–Al2024 composites prepared by mechanical processing in a high energy ball mill. J. Alloy. Compd. **536**, S26–S30 (2012)
23. M. Tavoosi, F. Karimzadeh, M.H. Enayati, Fabrication of Al-Zn/α-Al$_2$O$_3$ nanocomposite by mechanical alloying. Mater. Lett. **62**, 282–285 (2008)
24. L. Kollo, M. Leparoux, C.R. Bradbury, C. Jäggi, E. Carreño-Morelli, M. Rodríguez-Arbaizar, Investigation of planetary milling for nano-silicon carbide reinforced aluminium metal matrix composites. J. Alloy. Compd. **489**, 394–400 (2010)
25. L. Lu, M.O. Lai, J.L. Yeo, In situ synthesis of TiC composite for structural application. Compos. Struct. **47**, 613–618 (1999)
26. R. Derakhshandeh, H.A. Jenabali Jahromi, An investigation on the capability of equal channel angular pressing for consolidation of aluminum and aluminum composite powder. Mater. Des. **32**, 3377–3388 (2011)
27. M. De Cicco, L. Turng, X. Li, J.H. Perepezko, Nucleation catalysis in Aluminum alloy A356 using nanoscale inoculants. Metall. Mater. Trans. A **42**, 2323–2330 (2011)
28. M. De Cicco, L. Turng, X. Li, J.H. Perepezko, Production of semi-solid slurry through heterogeneous nucleation in metal matrix nanocomposites (MMNC) using nano-scaled ultrasonically dispersed inoculants. Solid State Phenom **141–143**, 487–492 (2008)
29. J. Li, W. Xu, X. Wu, H. Ding, K. Xia, Effects of grain size on compressive behaviour in ultrafine grained pure Mg processed by equal-channel angular pressing at room temperature. Mater. Sci. Eng. A **528**, 5993–5998 (2011)
30. A. Erman, J. Groza, X. Li, H. Choi, G. Cao, Nanoparticle effects in cast Mg-1 wt% SiC nano-composites. Mater. Sci. Eng. A **558**, 39–43 (2012)
31. Z. Wang, X. Wang, Y. Zhao, W. Du, SiC nanoparticles reinforced magnesium matrix composites fabricated by ultrasonic method. Trans. Nonferrous Met. Soc. China **20**, s1029–s1032 (2010)
32. H. Frenkel, B.L. Mordike, Magnesium strengthened by SiC nanoparticles. Mater. Sci. Eng. A **298**, 193–199 (2001)

33. K.B. Nie, X.J. Wang, L. Xu, K. Wu, X.S. Hu, M.Y. Zheng, Influence of extrusion temperature and process parameter on microstructures and tensile properties of a particulate reinforced magnesium matrix nanocomposites. Mater. Des. **36**, 199–205 (2012)
34. G. Cao, H. Choi, J. Oportus, H. Konishi, X. Li, Study on tensile properties and microstructure of cast AZ91D/AlN nanocomposites. Mater. Sci. Eng. A **494**, 127–131 (2008)
35. Q. Li, C.A. Rottmair, R.F. Singer, CNT reinforced light metal composites produced by melt stirring and by high pressure die casting. Compos. Sci. Technol. **70**, 2242–2247 (2010)
36. M.O. Lai, L. Lu, W. Laing, Formation of magnesium nanocomposite via mechanical milling. Compos. Struct. **66**, 301–304 (2004)
37. Kwangmin Choi, Jiyeon Seo, Donghyun Bae, Hyunjoo Choi, Mechanical properties of aluminum-based nanocomposite reinforced with fullerenes. Trans. Nonferrous Met. Soc. China **24**, s47–s52 (2014)
38. P. Luo, D.T. McDonald, W. Xu, S. Palanisamy, M.S. Dargusch, K. Xia, A modified Hall-Petch relationship in ultrafine-grained titanium recycled from chips by equal channel angular pressing. Scripta Mater. **66**, 785–788 (2012)
39. P. Luo, D.T. McDonald, S.M. Zhu, S. Palanisamy, M.S. Dargusch, K. Xia, Analysis of microstructure and strengthening in pure titanium recycled from machining chips by equal-channel angular pressing using electron backscatter diffraction. Mater. Sci. Eng. A **538**, 252–258 (2012)
40. V.V. Stolyarov, Y.T. Zhu, I.V. Alexandrov, T.C. Lowe, R.Z. Valiev, Influence of ECAP routes on the microstructure and properties of pure Ti. Mater. Sci. Eng. A **299**, 59–67 (2001)
41. S.M. Uddin, T. Mahmud, C. Wolf, C. Glanz, I. Kolaric, C. Volkmer, H. Höller, U. Wienecke, S. Roth, H. Fecht, Effect of size and shape of metal particles to improve hardness and electrical properties of carbon nanotube reinforced copper and copper alloy composites. Compos. Sci. Technol. **70**, 2253–2257 (2010)
42. D. Bozic, J. Stasic, B. Dimcic, M. Vilotijevic, V. Rajkovic, Multiple strengthening mechanisms in nanoparticle-reinforced copper matrix composites. J. Mater. Sci. **34**, 217–226 (2011)
43. J. Naser, W. Riehemann, H. Frenkel, Dispersion hardening of metals by nanoscaled ceramic powders. Mater. Sci. Eng. A **234–236**, 467–469 (1997)
44. P. Quang, Y.G. Jeong, S.C. Yoon, S.H. Hong, H.S. Kim, Consolidation of 1 vol.% carbon nanotube reinforced metal matrix nanocomposites via equal channel angular pressing. J. Mater. Process. Technol. **187–188**, 318–320 (2007)
45. H. Ahamed, V. Senthilkumar, Consolidation behavior of mechanically alloyed aluminum based nanocomposites reinforced with nanoscaled Y_2O_3/Al_2O_3. Mater. Charact. **62**, 1235–1249 (2011)
46. S.R. Bakshi, D. Lahiri, A. Agarwal, Carbon nanotube reinforced metal composites-a review. Int. Mater. Rev. **55**, 42–64 (2010)
47. S.S. Nayak, S.K. Pabi, D.H. Kim, B.S. Murty, Microstructure-hardness relationship of Al–$(L1_2)Al_3Ti$ nanocomposites prepared by rapid solidification processing. Intermetallics **18**, 487–492 (2010)
48. E.O. Hall, The deformation and aging of mild steel. Proc. Phys. Soc. London, Sec. B **64**, 747–753 (1951)
49. N.J. Petch, The cleavage strength of polycrystals. J. Iron Steel Res. **174**, 25–28 (1953)
50. D. Hull, D.J. Bacon, *Introduction to Dislocations*, 4th edn. (Butterworth-Heinemann, London, 2001)
51. D. Hull, T.W. Clyne, An Introduction to Composite Materials. Cambridge Solid State Science Series. 2th edn. (1996)
52. A. Sanaty-Zadeh, Comparison between current models for the strength of particulate-reinforced metal matrix nanocomposites with emphasis on consideration of Hall-Petch effect. Mater. Sci. Eng. A **531**, 112–118 (2012)
53. Z. Zhang, D.L. Chen, Contribution of Orowan strengthening effect in particulate-reinforced metal matrix nanocomposites. Mater. Sci. Eng. A **483–484**, 148–152 (2008)

54. Z. Zhang, D.L. Chen, Consideration of Orowan strengthening effect in particulate-reinforced metal matrix nanocomposites: a model for predicting their yield strength. Scripta Mater. **54**, 1321–1326 (2006)
55. A.H. Chokski, A. Rosen, J. Karch, H. Gleiter, On the validity of the Hall–Petch relationship in nanocrystalline materials'. Scr. Metall. **23**, 1679–1683 (1989)
56. G.W. Nieman, J.R. Weertman, R.W. Siegel, Microhardness of nanocrystallinepalladium and copper produced by inert-gas condensation. Scr. Metall. **23**, 2013–2018 (1989)
57. K. Lu, W.D. Wei, J.T. Wang, Microhardness and fracture properties of nanocrystalline Ni–P alloy. Scr. Metall. Mater. **24**, 2319–2323 (1990)
58. G.W. Nieman, J.R. Weertman, R.W. Siegel, Mechanical behavior of nanocrystalline Cu and Pd. J. Mater. Res. **6**, 1012–1027 (1991)
59. G.E. Fougere, J.R. Weertman, R.W. Siegel, S. Kim, Grain-size dependent hardening and softening of nanocrystalline Cu and Pd. Scr. Metall. Mater. **26**, 1879–1883 (1992)
60. A.M. El-Sherik, U. Erb, G. Palumbo, K.T. Aust, Deviations from Hall-Petch behavior in as-prepared nanocrystalline nickel. Scr. Metall. Mater. **27**, 1185–1188 (1992)
61. V.Y. Gertsman, M. Hoffmann, H. Gleiter, R. Dirringer, The study of grain size dependence of yield of copper. Acta Metall. Mater. **42**, 3539–3544 (1994)
62. P.G. Sanders, J.A. Eastman, J.R. Weertman, Elastic and tensile behavior of nanocrystalline copper and palladium. Acta Mater. **45**, 4019–4025 (1997)
63. N. Wang, Z. Wang, K.T. Aust, U. Erb, Room temperature creep behavior of nanocrystalline nickel produced by an electrodeposition technique. Mater. Sci. Eng. A. **237**, 150–158 (1997)
64. C.A. Schuh, T.G. Nieh, T. Yamasaki, Hall-Petch breakdown manifested in abrasive wear resistance of nanocrystalline nickel. Scripta Mater. **46**, 735–740 (2002)
65. A. Giga, Y. Kimoto, Y. Takigawa, K. Higashi, Demonstration of an inverse Hall-Petch relationship in electrodeposited nanocrystalline Ni–W alloys through tensile testing. Scripta Mater. **55**, 143–146 (2006)
66. C.E. Carlton, P.J. Ferreira, What is Behind the Inverse Hall-Petch effect in nanocrystalline materials? Acta Mater. **55**, 3749–3756 (2007)
67. Z.Y. Ma, Y.L. Li, Y. Liang, F. Zheng, J. Bi, S.C. Tjong, Nanometric Si_3N_4 particulate-reinforced aluminum composite. Mater. Sci. Eng. A **219**, 229–231 (1996)
68. D.J. Lloyd, Particle-reinforced aluminum and magnesium matrix composites. Int. Mater. Rev. **39**, 1–23 (1994)
69. N.A. Fleck, M.F. Ashby, J.W. Hutchinson, The role of geometrically necessary dislocations ingiving material strengthening. Scripta Mater. **48**, 179–183 (2003)
70. R.J. Arsenault, N. Shi, Dislocation generation due to differences between the coefficients of thermal-expansion. Mater. Sci. Eng. A **81**, 175–187 (1986)
71. R.E. Smallman, A.H.W. Ngan, Physical Metallurgy and Advanced Materials, 7th edn (Butterworth-Heinemann, London, 2007)
72. M.A. Azouni, P. Casses, Thermophysical properties effects on segregation during solidification. Adv. Colloid Interface Sci. **75**, 83–106 (1998)
73. J. Lan, Y. Yang, X. Li, Microstructure and microhardness of SiC nanoparticles reinforced magnesium composites. Mater. Sci. Eng. A **386**, 284–290 (2004)
74. G. Cao, J. Kobliska, H. Konishi, X. Li, Tensile properties and microstructure of SiC nanoparticle-reinforced Mg-4Zn alloy fabricated by ultrasonic cavitation-based solidification processing. Metall. Mater. Trans. A **39**, 880–886 (2008)
75. Chapter Insight into Designing Biocompatible Magnesium Alloys and Composites Part of the series SpringerBriefs in Materials, pp. 17–34 Date: 15 January 2015 Synthesis of Magnesium-Based Biomaterials
76. N. Srikanth, K.F. Ho, M. Gupta, Effect of length scale of alumina particles of different sizes on the damping characteristics of an Al–Mg alloy. Mater. Sci. Eng. A **423**, 189–191 (2006)
77. D. Gu, Y.C. Hagedoorn, W. Meiners, K. Wissenbach, R. Poprawe, Nanocrystalline TiC reinforced Ti matrix bulk-form nanocomposites by selective laser melting (SLM): densification, growth mechanism and wear behavior. Compos. Sci. Technol. **71**, 1612–1620 (2011)

78. J.P. Tu, N.Y. Wang, Y.Z. Yang, W.X. Qi, F. Liu, X.B. Zhang, H.M. Lu, M.S. Liu, Preparation and properties of TiB_2 nanoparticles reinforced copper matrix composites by in situ processing. Mater. Lett. **52**, 448–452 (2002)

79. N.L. Yue, L. Lu, M.O. Lai, Application of thermodynamic calculation in the in-situ process of Al/TiB_2. Compos. Struct. **47**, 691–694 (1999)

80. Y.Q. Liu, H.T. Cong, W. Wang, C.H. Sun, H.M. Cheng, AlN nanoparticle-reinforced nanocrystalline Al matrix composites: fabrication and mechanical properties. Mater. Sci. Eng. A **505**, 151–156 (2009)

81. Z. Bian, M.X. Pan, Y. Zhang, W.H. Wang, Carbon-nanotube-reinforced $Zr_{52.5}Cu_{17.9}$ $Ni_{14.6}Al_{10}Ti_5$ bulk metallic glass composites. Appl. Phys. Lett. **81**, 4739–4741 (2002)

82. Z. Bian, R.J. Wang, W.H. Wang, T. Zhang, A. Inoue, Carbon-nanotube-reinforced Zr-based bulk metallic glass composites and their properties. Adv. Funct. Mater. **14**, 55–63 (2004)

83. C.S. Goh, J. Wei, L.C. Lee, M. Gupta, Ductility improvement and fatigue studies in Mg-CNT nanocomposites. Compos. Sci. Technol. **68**, 1432–1439 (2008)

84. E. El-Kady, T. Mahmoud, A. Ali, On the electrical and thermal conductivities of Cast A356/Al_2O_3 metal matrix nanocomposites. Mater. Sci. Appl. **22**, 1180–1187 (2011)

85. M. De Cicco, X. Li, L.S. Turng, Semi-solid casting (SSC) of zinc alloy nanocomposites. J. Mater. Process. Technol. **209**, 5881–5885 (2009)

86. L. Lu, M.O. Lai, Formation of new materials in the solid state by mechanical alloying. Mater. Des. **16**, 33–39 (1995)

87. F. Zhang, W.A. Kacmarek, L. Lu, M.O. Lai, Formation of Al TiN metal matrix composite via mechanochemical route. Scripta Mater. **43**, 1097–1102 (2000)

88. M. Mozaffari, M. Gheisari, M. Niyaifar, J. Amighian, Magnetic properties of mechanochemically prepared iron-wustite (Fe-FeyO) nanocomposites. J. Magn. Magn. Mater. **321**, 2981–2984 (2009)

89. Z. Razavi Hesabi, A. Simchi, S.M. Seyed Reihani, Structural evolution during mechanical milling of nanometric and micrometric Al_2O_3 reinforced Al matrix composites. Mater. Sci. Eng. A **428**, 159–168 (2006)

90. L. Lu, M.O. Lai, Y.H. Toh, L. Froyen, Structure and properties of Mg-Al-Ti-B alloys synthesized via mechanical alloying. Mater. Sci. Eng. A **334**, 163–172 (2002)

91. E. Mostaed, H. Saghafian, A. Mostaed, A. Shokuhfar, H.R. Rezaie, Investigation on preparation af Al-4.5 %Cu/SiCp nanocomposites powder via mechanical milling. Powder Technol. **221**, 278–283 (2012)

92. L. Lu, M.O. Lai, W. Liang, Magnesium nanocomposites via mechanochemical milling. Compos. Sci. Technol. **64**, 2009–2014 (2004)

93. M.A. Thein, L. Lu, M.O. Lai, Effect of milling and reinforcement on mechanical properties of nanostructured magnesium composite. J. Mater. Process. Technol. **209**, 4439–4443 (2009)

94. R. Casati, Q. Ge, M. Vedani, D. Dellasega, P. Bassani, A. Tuissi, Preparazione di nano-compositi a matrice metallica Al/Al_2O_3 mediante ECAP e estrusione a caldo. La Metallurgia Italiana **105**, 25–30 (2013)

95. F. Shehata, A. Fathy, M. Abdelhameed, S.F. Mustafa, Preparation and properties of Al_2O_3 nanoparticle reinforced copper matrix composites by in situ processing. Mater. Des. **30**, 2756–2762 (2009)

96. Z. Trojanova, P. Lukac, H. Ferkel, W. Riehemann, Internal friction in microcrystalline and nanocrystalline Mg. Mater. Sci. Eng. A **370**, 154–157 (2004)

97. R. Casati, F. Bonollo, D. Dellasega, A. Fabrizi, G. Timelli, A. Tuissi, M. Vedani, Ex situ Al-Al_2O_3 ultrafine grained nanocomposites produced via powder metallurgy. J. Alloy. Compd. **615**, S386–S388 (2014)

98. R. Casati, M. Amadio, C.A. Biffi, D. Dellasega, A. Tuissi, M. Vedani, Al/Al_2O_3 nano-composite produced by ECAP. Mater. Sci. Forum. **762**, 457–464 (2013)

99. X. Xia, Consolidation of particles by severe plastic deformation: mechanism and applications in processing bulk ultrafine and nanostructured alloys and composites. Adv. Eng. Mater. **12**, 724–729 (2010)

100. R. Derakhshande, H.S.A. Jenabali Jahromi, B. Esfandiar. Simulation aluminum powder in tube compaction using equal channel angular pressing. J. Mater. Eng. Perform. **21**, 143–152 (2012)

101. C. Suryanarayana, Mechanical alloying and milling. Prog. Mater Sci. **46**, 1–184 (2001)

102. S.C. Tjong, H. Chen, Nanocrystalline materials and coatings. Mater. Sci. Eng. R. **45**, 1–88 (2004)

103. J. Benjamin, Mechanical alloying. Sci. Am. **234**, 40–49 (1976)

104. P.Y. Lee, J.L. Yang, H.M. Lin, Amorphization behaviour in mechanically alloyed Ni—Ta powders. J. Mater. Sci. **33**, 235–239 (1998)

105. J.S. Benjamin, T.E. Volin, Mechanism of mechanical alloying. Metall. Trans. **5**, 1929–1934 (1974)

106. P.S. Gilman, J.S. Benjamin, Mechanical alloying. Annu. Rev. Mater. Sci. **13**, 279–300 (1983)

107. J.S. Benjamin, Mechanical alloying—A perspective. Met. Powder Rep. **45**, 122–127 (1990)

108. C. Suryanarayana, E. Ivanov, R. Noun, M.A. Contreras, J.J. Moore, Phase selection in a mechanically alloyed Cu-In-Ga-Se powder mixture. J. Mater. Res. **14**, 377–383 (1999)

109. B.L. Chu, C.C. Chen, T.P. Perng, Amorphization of Ti_{1-x} Mn_x. Metall. Trans. A. Phys. Metall. Mater. Sci. **23**, 2105–2110 (1992)

110. K. Tokumitsu, Synthesis of metastable Fe_3C, Co_3C and Ni_3C by mechanical alloying method. Mater. Sci. Forum **235–238**, 127–132 (1997)

111. B.K. Yen, T. Aizawa, J. Kihara, Synthesis and formation mechanisms of molybdenum silicides by mechanical alloying. Mater. Sci. Eng. A **220**, 8–14 (1996)

112. B.K. Yen, T. Aizawa, J. Kihara, Mechanical alloying behavior in molibdenum-silicon system. Mater. Sci. Forum. **235–238,** 157–162 (1997)

113. A. Tonejc, D. Duzevic, A.M. Tonejc, Effects of ball milling on pure antimony, on Ga—Sb alloy and on Ga + Sb powder mixture; oxidation, glass formation and crystallization. Sci. Eng. A **134**, 1372–1375 (1991)

114. T. Ohtani, K. Maruyama, K. Ohshima, Synthesis of copper, silver, and samarium chalcogenides by mechanical alloying. Mater. Res. Bull. **32**, 343–350 (1997)

115. W. Guo, A. Iasonna, M. Magini, S. Martelli, F. Padella, Synthesis of amorphous and metastable $Ti_{40}Al_{60}$ alloys by mechanical alloying of elemental powders. J. Mater. Sci. **29**, 2436–2444 (1994)

116. F. Padella, E. Paradiso, N. Burgio, M. Magini, S. Martelli, W. Guo, A. Iasonna, Mechanical alloying of the Pd-Si system in controlled conditions of energy transfer. J. Less Common Metals **175**, 79–90 (1991)

117. K.B. Gerasimov, A.A. Gusev, E.Y. Ivanov, V.V. Boldyrev, Tribochemical equilibrium in mechanical alloying of metals. J. Mater. Sci. **26**, 2495–2500 (1991)

118. L. Liu, S. Casadio, M. Magini, C.A. Nannetti, Y. Qin, K. Zheng, Solid state reactions of $V_{75}Si_{25}$ driven by mechanical alloying. Mater. Sci. Forum **235–238**, 163–168 (1997)

119. L. Takacs, M. Pardavi-Horvath, Nanocomposite formation in the Fe_3O_4-Zn system by reaction milling. J. Appl. Phys. **75**, 5864–5866 (1994)

120. W.E. Frazier, M.J. Koczak, Mechanical and thermal stability of powder metallurgy aluminum-titanium alloys. Scr. Metall. **21**, 129–134 (1987)

121. G. Chen, K. Wang, J. Wang, H. Jiang, M. Quan. In: deBarbadillo, J.J. et al. (Eds). Mechanical alloying for structural applications. ASM International, Materials Park, OH, 1993) pp. 183–187

122. P.K. Ivison, N. Cowlam, I. Soletta, G. Cocco, S. Enzo, L. Battezzati, The influence of hydrogen contamination on the amorphization reaction of CuTi alloys. Mater. Sci. Eng. A **134**, 859–862 (1991)

123. P.K. Ivison, I. Soletta, N. Cowlam, G. Cocco, S. Enzo, L. Battezzati, Evidence of chemical short-range order in amorphous CuTi alloys produced by mechanical alloying. J. Phys.: Condens. Matter. **4**, 1635–1645 (1992)

124. L.B. Hong, C. Bansal, B. Fultz, Steady state grain size and thermal stability of nanophase Ni$_3$Fe and Fe$_3$X (X=Si, Zn, Sn) synthesized by ball milling at elevated temperatures. Nanostruct. Mater. **4**, 949–956 (1994)

125. C.C. Koch, D. Pathak, K. Yamada. In: deBarbadillo JJ, et al. (Eds) Mechanical alloying for structural applications. (ASM International, Materials Park, OH, 1993), pp. 205–212

126. H. Kimura, M. Kimura, in *Solid State Powder Processing*, ed. by A.H. Clauer, J. J. deBarbadillo (TMS, Warrendale, PA, 1990), pp. 365–377

127. C.H. Lee, M. Mori, T. Fukunaga, U. Mizutani, Effect of ambient temperature on the MA and MG processes in Ni–Zr alloy system. Jpn. J. Appl. Phys. **29**, 540–544 (1990)

128. V.M. Segal, Methods of stress-strain analysis in metalforming. (Physical Technical Institute Academy of Sciences of Buelorussia, Minsk, Russia, 1974)

129. R.Z. Valiev, N.A. Krasilnikov, N.K. Tsenev, Plastic-deformation of alloys with submicron-grained structure. Mater. Sci. Eng. A **137**, 35–40 (1991)

130. R.Z. Valiev, A.V. Korznikov, R.R. Mulyukov, Structure and properties of ultrafine-grained materials produced by severe plastic-deformation. Mater. Sci. Eng. A **168**, 141–148 (1993)

131. V.M. Segal, Engineering and commercialization of equal channel angular extrusion (ECAE). Mater. Sci. Eng. A **386**, 269–276 (2004)

132. R.Z. Valiev, R.K. Islamgaliev, I.V. Alexandrov, Bulk nanostructured materials from severe plastic deformation. Prog. Mater Sci. **45**, 103–189 (2000)

133. Y. Iwahashi, Z. Horita, M. Nemoto, T.G. Langdon, An investigation of microstructural evolution during equal-channel angular pressing. Acta Mater. **45**, 4733–4741 (1997)

134. Y. Iwahashi, J. Wang, Z. Horita, M. Nemoto, T.G. Langdon, Principle of equal-channel angular pressing for the processing of ultra-fine grained materials. Scripta Mater. **35**, 143–146 (1996)

135. V.M. Segal, Equal channel angular extrusion: from macromechanics to structure formation. Mater. Sci. Eng. A **271**, 322–333 (1999)

136. R.Z. Valiev, T.G. Langdon, Principles of equal-channel angular pressing as a processing tool for grain refinement. Prog. Mater Sci. **51**, 881–981 (2006)

137. Y. Iwahashi, Z. Horita, M. Nemoto, T.G. Langdon, The process of grain refinement in equal-channel angular pressing. Acta Mater. **46**, 3317–3331 (1998)

138. T.G. Langdon, The characteristics of grain refinement in materials processed by severe plastic deformation. Rev. Adv. Mater. Sci. **13**, 6–14 (2006)

139. R.Z. Valiev, T.G. Langdon, Developments in the use of ECAP processing for grain refinement. Rev. Adv. Mater. Sci. **13**, 15–26 (2006)

140. R.Z. Valiev, R.K. Islamgaliev, N.F. Kuzmina, T.G. Langdon, Strengthening and grain refinement in an Al-6061 metal matrix composite through intense plastic straining. Scripta Mater. **40**, 117–122 (1998)

141. I. Gutierrez-Urrutia, M.A. Munoz-Morris, D.G. Morris, The effect of coarse second-phase particles and fine precipitates on microstructure refinement and mechanical properties of severely deformed Al alloy. Mater. Sci. Eng. A **394**, 399–410 (2005)

142. K. Nakashimaa, Z. Horitaa, M. Nemoto, T.G. Langdon, Development of a multi-pass facility for equal-channel angular pressing to high total strains. Mater. Sci. Eng. A **281**, 82–87 (2000)

143. A. Yamashita, D. Yamaguchia, Z. Horitaa, T.G. Langdonb, Influence of pressing temperature on microstructural development in equal-channel angular pressing. Mater. Sci. Eng. A **287**, 100–106 (2000)

144. D.H. Shin, J.J. Pak, Y.K. Kim, K.T. Park, Y.S. Kim, Effect of pressing temperature on microstructure and tensile behavior of low carbon steels processed by equal channel angular pressing. Mater. Sci. Eng. A **325**, 31–37 (2002)

145. Y.C. Chen, Y.Y. Huang, C.P. Chang, P.W. Kao, The effect of extrusion temperature on the development of deformation microstructures in 5052 aluminium alloy processed by equal channel angular extrusion. Acta Mater. **51**, 2005–2015 (2003)

146. W.H. Huang, C.Y. Yu, P.W. Kao, C.P. Chang, The effect of strain path and temperature on the microstructure developed in copper processed by ECAE. Mater. Sci. Eng. A **366**, 221–228 (2004)

147. K. Xia, X. Wu, T. Honma, S.P. Ringer, Ultrafine pure aluminium through back pressure equal channel angular consolidation (BP-ECAC) of particles. J. Mater. Sci. **42**, 1551–1560 (2007)

148. W. Xu, T. Honma, X. Wu, S.P. Ringer, High strength ultrafine/nanostructured aluminum produced by back pressure equal channel angular processing. Appl. Phys. Lett. **91**, 031901 (2007)

149. A. Bohner, F. Kriebel, R. Lapovok, H.W. Höppel, M. Göken, Influence of backpressure during ECAP on the monotonic and cyclic deformation behavior of Aa_{5754} and $Cu_{99.5}$. Adv. Eng. Mater. **13**, 269–274 (2011)

150. R.Y. Lapovok, The role of back-pressure in equal channel angular extrusion. J. Mater. Sci. **40**, 341–346 (2005)

151. P.W.J. McKenzie, R. Lapovok, Y. Estrin, The influence of back pressure on ECAP processed. AA 6016: modeling and experiment. Acta Mater. **55**, 2985–2993 (2007)

152. V.V. Stolyarov, R. Lapovok, Effect of backpressure on structure and properties of AA5083 alloy processed by ECAP. J. Alloy. Compd. **378**, 233–236 (2004)

153. V.V. Stolyarov, R. Lapovokb, I.G. Brodovac, P.F. Thomsonb, Ultrafine-grained Al-5 wt.% Fe alloy processed by ECAP with backpressure. Mater. Sci. Eng. A **357**, 159–167 (2003)

154. J.Z. Li, W. Xu, X. Wu, H. Ding, K. Xia, Effects of grain size on compressive behaviour in ultrafine grained pure Mg processed by equal channel angular pressing at room temperature. Mater. Sci. Eng. A **528**, 5993–5998 (2011)

155. C. Xu, K. Xia, T.G. Langdon, Processing of a magnesium alloy by equal-channel angular pressing using a back-pressure. Mater. Sci. Eng., A **527**, 205–211 (2009)

156. S. Xiang, K. Matsuki, N. Takatsuji, M. Tokizawa, T. Yokote, J. Kusui, K. Yokoe, Microstructure and mechanical properties of PM 2024Al-3Fe-5Ni alloy consolidated by a new process, equal channel angular pressing. J. Mater. Sci. Lett. **16**, 1725–1727 (1997)

157. R. Lapovok, D. Tomus, C. Bettles, Shear deformation with imposed hydrostatic pressure for enhanced compaction of powder. Scripta Mater. **58**, 898–901 (2008)

158. W. Xu, X. Wu, D. Sadedin, G. Wellwood, K. Xia, Ultrafine-grained titanium of high interstitial contents with a good combination of strength and ductility. Appl. Phys. Lett. **92**, 0119241-3 (2008)

159. T. Sheppard, Extrusion of aluminium alloys. Kluwer Academic Press, Boston USA. Dordrecht. The Netherlands (1999), ISBN 041259070 0

160. P. Loewenstein, L.R. Aronin, A.L. Geary, Powder Metallurgy, in W. Leszyski, (ed.), (Interscience, 1961) pp. 563–583

161. Powder Metal Technologies and Applications was published in 1998 as Volume 7 of ASM Handbook

162. P. Roberts, Technical Paper MF 76-391, SME (1976)

163. R. Casati, A. Fabrizi, A. Tuissi, K. Xia, M. Vedani, ECAP consolidation of Al matrix composites reinforced with in-situ $\gamma\text{-}Al_2O_3$ nanoparticles. Mater. Sci. Eng. A **648**, 113–122 (2015)

Chapter 2
Experimental Methods

Abstract The experimental work was mainly aimed at producing Al based nanocomposites reinforced with alumina NPs. In order to attain a homogeneously dispersion of nanoparticles, several powder metallurgy routes were adopted. They relied on high-energy ball milling and powder compaction either via ECAP or hot extrusion. Different processing parameters were used in order to break the clusters of NPs and to improve the final properties of the composites. The powders and the consolidated materials were analyzed from the morphological, microstructural and mechanical point of view. Several characterization techniques were used to this purpose and described in this chapter.

Keywords Powder metallurgy · ECAP · Extrusion · Ball milling · Metal matrix nanocomposites · Aluminum · Alumina nanoparticles · Mechanical tests · Instrumented indentation · Electron microscopy

2.1 Materials

2.1.1 Al Powders

The following commercial Al powders were used as starting materials:

- commercial purity (CP) Al powder with maximum size lower than 45 μm (supplied by ECKA Granules Germany and ECKA Granules Australia);
- CP Al nanoparticles with maximum size lower than 80 nm and BET = 23.4 m^2/g.

The nominal oxygen content in the micro-Al powder was estimated to be lower than 0.5 wt% (data supplied by ECKA Granules), the one of nano-Al powder was lower than 10 wt%. The Al powder particles are indeed covered by a 2–4 nm thick oxide layer [1–6].

© The Author(s) 2016
R. Casati, *Aluminum Matrix Composites Reinforced with Alumina Nanoparticles*,
PoliMI SpringerBriefs, DOI 10.1007/978-3-319-27732-5_2

2.1.2 Al_2O_3 Nanoparticles

The following commercial alumina nanoparticles were used as reinforcing phase for preparation of nanocomposites:

- colloidal solution of alumina particles with maximum size lower than 50 nm in isopropyl alcohol (supplied by Sigma Aldrich);
- dry nano-sized particles of Al_2O_3 with an average particle size of 20 nm (supplied by Cometox).

2.2 Processing

2.2.1 High-Energy Ball Milling

High-energy ball milling was performed by two planetary mills, the schematic view of the movement of the main disk and of the planets is depicted in Fig. 2.1. A Vario-Planetary Mill Pulverisette 4 and a QM-1SP4 equipped with steel vials and balls (10 mm in diameter) were used to mill pure Al and Al-Al_2O_3 powders. To avoid excessive cold welding and agglomeration of the particles, different types and amount of PCAs were used in the experiments, namely: 10 wt% of ethanol, 1.5 wt% of ethanol and 1.5 wt% of stearic acid ($C_{18}H_{36}O_2$). To minimize oxidation during milling, the two vials were back-filled with argon. The milling was performed for different times (namely: 2, 5, 16 and 24 h) with a ball-to powder weight ratio of 5:1 and 10:1. An excessive temperature rise was avoided by interrupting the procedure for 10 min after each 30 min of milling. With the Pulverisette 4, the speed of the main disk was set at 250 rpm clockwise whereas the speed of the two planets at 200 rpm counter-clockwise. With the QM-1SP4, the main speed was set at 300 rpm, while that of the planets was set at 100 rpm.

Fig. 2.1 Schematic of the movement of the main disk and of the vials into the planetary mill

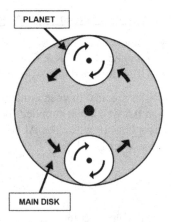

2.2.2 Equal Channel Angular Pressing (ECAP)

The ECAP die consisted of two cylindrical channels, an entrance channel and an exit channel, of equal cross sectional dimensions (diameter = 10 mm) intersecting at an angle of $\Phi = 110°$. The equivalent strain per pass is $\varepsilon_e = 0.76$ per pass. The ECAP die can be separated into two halves to facilitate the cleaning procedure of the channels and the removal of accidentally stuck samples. It can also be heated by 4 resistances (800 W each) to perform high temperature pressing.

To perform ECAP, the powders (milled or as-received) were packed in cylindrical Cu alloy or Al alloy cans. They were closed by means of press-fit plugs and then subjected to ECAP at different temperatures ranging from 200 to 400 °C. The plunger speed was set to 5 mm/min and the temperature was monitored by a type-K thermocouple. Molikote MoS_2 was used as lubricant. The samples always went through the route B_c during multipass pressing.

2.2.2.1 Back-Pressure Equal Channel Angular Pressing (BP-ECAP)

BP-ECAP was also used to consolidate pure Al micro and nano powder and a mix of them. The die consisted of two cylindrical channels of equal cross sectional dimensions (diameter = 11 mm) and intersection angle of 90°. The equivalent strain per pass is $\varepsilon_e = 1.03$. The ECAP die can be separated into two halves and it can also be heated by 2 heating blankets. In this case, loose powder was wrapped in steel foils and cold compacted before being placed into the channel. The channel wall and the plungers were lubricated with graphite sheet to minimize friction. Two pistons of up to 25 tons in capacity provided the forward force through the plunger in the entrance channel and the back pressure through the back plunger in the exit channel. The pistons were set to move at a constant speed of 5 mm/min. After the first pass, the forward and backward plungers swapped roles and the specimen was pressed back through the shearing plane into the original channel. By repeatedly pressing the specimen forward and backward, the specimen is subjected to shearing patterns equivalent to those experienced in route C.

2.2.3 Hot Extrusion

The powders were packed in cylindrical Cu alloy cans with external diameter of 10 mm and thickness of 1 mm. They were closed by means of press-fit plugs and then subjected to hot extrusion at 400 °C. The extrusion speed was 5 mm/min. The die was heated by an induction coil and the temperature was monitored by a type-K thermocouple. The starting billet diameter was reduced to 4 mm after extrusion (R = 6.25).

2.2.4   Rolling

After powder consolidation through hot extrusion, the can was stripped off and the exposed nanocomposite rods were then cold rolled down to a square section of 1 mm^2 by a caliber rolling mill, with intermediate annealing at 400 °C for 5 min after each area reduction of about 20 %.

2.3   Characterization Techniques

2.3.1   Scanning Electron Microscope (SEM)

Microstructural analysis was performed using the following scanning electron microscopes (SEM) with conventional thermo-ionic electron gun or field-emission gun (FEG):

- FEG-SEM FEI Quanta 200F (available at the University of Melbourne);
- FEG-SEM Zeiss Supra 40 (available at Politecnico di Milano, Department of Energy);
- FEG-SEM FEI Quanta 250 (available at the University of Padova);
- SEM Zeiss EVO 50 (available at Politecnico di Milano, Department of Mechanical Engineering).

SEM analysis was performed on powdered and bulk samples using secondary electron and back-scattered electron detectors. The samples were mounted in resin and polished following a standard metallography procedure. To obtain better contrast between the NPs and matrix, the samples were etched using the Keller's solution.

For some samples, the mean Al matrix grain size and the distribution of grain boundary (GB) misorientation were obtained by using electron backscatter diffraction (EBSD). The EBSD analysis was carried out on billet sections cut normally to pressing direction (i.e. along axial direction) in the center of the samples. Before the measurements, the surface of the specimens was polished by low-angle ion milling. The GB misorientation distribution of samples was calculated by considering misorientation angles (θ) larger than 2°. GBs with θ between 2° and 15° have been defined as low-angle grain boundaries (LAGBs), whereas those with θ exceeding 15° have been referred as high-angle grain boundaries (HAGBs).

2.3.2   Transmission Electron Microscopy (TEM)

TEM foils with a final thickness of 100 nm were prepared from rolled wires using a Nova Nanolab 200 focused ion beam (FIB). In Fig. 2.2, a picture captured during the cutting process is shown. High angle annular dark field (HAADF) imaging by

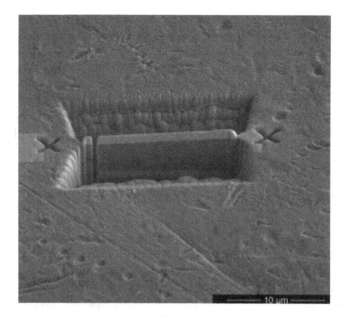

Fig. 2.2 Nanocomposite slice being cut by FIB [7]

scanning transmission electron microscopy (STEM) was performed using a FEI Tecnai F20 operating at 200 kV available at the University of Melbourne.

Discs with 3-mm diameter for TEM investigations were transversally cut from the ECAP billets. They were mechanically ground to 30 μm thickness, and finally reduced by ion milling process. TEM inspections were performed using a JEOL JEM 2000 EX II operating at 200 kV available at the University of Padova.

2.3.3 X-Ray Diffraction (XRD)

A Philips Panalytical X'Pert diffractometer (Ni-filtered Cu Kα radiation with wavelength of 1.542 Å and a X'Celerator detector operating at 40 kV and 20 mA) was used to analyze bulk and powder samples. XRD patterns were recorded between 5° and 120° (2θ) at a step size of about 0.01° and scan time of 1 s. The crystallite size and micro-strains were calculated using the Williamson–Hall analysis [8] by means of the software X-Pert Highscore.

2.3.4 Tensile and Compression Testing

Tensile and compression tests were performed with MTS Alliance RT/100. Compressive specimens (4 mm in diameter and 6 mm in length) were cut from the samples consolidated by BP-ECAP with the compression axis parallel to the longitudinal direction. Tensile tests were indeed performed on nanocomposites consolidated via hot extrusion due to size limitations for other processing routes. For tensile and compression tests, a crosshead speed of 0.5 mm/min was adopted. Since the wire samples for tensile tests were too short to adopt special clamps for wires, premature fractures occasionally occurred close to the clamping position at strain levels exceeding about 4.5 %. For this reason, the stress–strain curves presented have been are terminated at a strain of 4.5 %. For compression tests, preload of 50 N and a crosshead speed of 0.5 mm/min was set.

2.3.5 Vickers Hardness

Vickers microhardness (HV) testing was performed on consolidated and powdered materials using a Future Tech Corp. FM-700. Before indentation, powder samples were mounted. Both powder and bulk nanocomposite surfaces were prepared by grinding and polishing. The load used for indentation was 200 g and the dwell time for each measurement was 15 s. Hardness for each sample was generally taken as the average value from at least 8 measurements.

2.3.6 Density Testing

Densities of consolidated samples were measured based on the Archimedes principle. After being polished, the specimens were first weighed in air and then in distilled water. The density was calculated using the following equation:

$$\rho = \frac{A}{A - B} \rho_0 \tag{2.1}$$

where A is the weight in air, B is the weight in distilled water and ρ_0 is the density of the distilled water at the testing temperature (i.e. room temperature). Five measurements were obtained from each sample to calculate the average density.

2.3.7 Instrumented Indentation

The tensile flow properties of the materials were estimated by instrumented micro-indentation cyclic tests. For this purpose, CSM Instruments Micro-Combi Tester was employed. 10 progressive linear mechanical cycles ranging from 2 to 20 N were performed (Fig. 2.3a) on polished surface of samples. The loading/unloading speed was set to 3 N/min while the peak load was held for 5 s after each loading ramp. Spherical diamond tip indenter with 200 μm radius was

Fig. 2.3 a Progressive linear
loading cycles from 2 to 20 N
used for micro-indentation
tests. **b** Cross section of an
indentation profile showing
the pile-up and sink-in
phenomena that can occur
during indentation. Ball
radius (R), contact depth (h$_C$),
pile up depth (s) and
indentation depth (h) are the
variables to consider for the
estimation of the real contact
area between the indenter and
the specimen [11, 12]

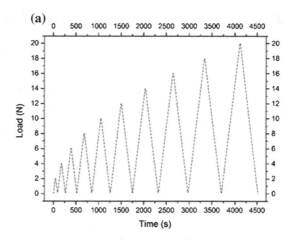

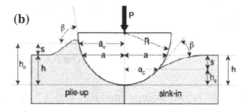

used. The stress/strain curve in the plastic regime was estimated by using the
method described in Ref. [9], where the strain was calculated as follow:

$$\varepsilon = \ln\frac{2}{1 + \cos\gamma} \qquad (2.2)$$

where γ is the contact angle between the indenter and the specimen; while the stress
was calculated as follow:

$$\sigma = \frac{P_m}{\Psi} \qquad (2.3)$$

Here, P_m is the mean contact pressure and Ψ is a coefficient, which is almost
constant in the plastic regime. In addition, the pile up (Fig. 2.3b) produced during
indentation was taken into account for the estimation of the real contact area
between the indenter and the specimen as reported in Ref. [10]. The strain-hardening
exponent (n) of the Hollomon's equation was then calculated form the obtained
data as the linear slope of the true stress versus true strain curves in log-log plots.

It is worth mentioning that the tests were performed according to the ISO/TR
29381 "Metallic materials—Measurement of mechanical properties by an instrumented indentation test—Indentation tensile properties" [11] which describes the
methods for evaluating tensile properties of metallic materials (true stress-strain
curve) using an instrumented indentation test.

2.3.8 Dynamic-Mechanical Analysis (DMA)

Mechanical spectroscopy [13] is a technique that consists of applying a sinusoidal stress to a material and measuring the strain response. The internal friction (IF) is related to the time-dependent elasticity of a material. Metals and alloys respond to an applied load not only by time-independent elastic strain, but also by time-dependent strain that lags behind the applied load. Because of the lag induced by the relaxation, the stress σ and strain ε can be expressed as

$$\sigma = \sigma_0 \exp(i\omega t) \tag{2.4}$$

$$\varepsilon = \varepsilon_0 \exp(i\omega t - \delta) \tag{2.5}$$

where σ_0 and ε_0 are the stress and strain amplitudes, respectively, x is the angular vibration frequency and d is the loss angle by which the strain lags behind the stress. By combining these two equations, the resultant complex modulus, E is defined as:

$$E = \frac{\sigma_0}{\varepsilon_0}(\cos\delta + i\sin\delta) = E' + iE'' \tag{2.6}$$

where E' is the storage modulus and E'' the loss modulus. The storage modulus represents the stiffness whereas the loss modulus is a measure of the oscillation energy transformed into heat. The ratio of the loss modulus to the storage modulus is usually defined as $\tan\delta$ and commonly used to indicate the damping capacity of a material:

$$\tan\delta = \frac{E''}{E'} \tag{2.7}$$

Accordingly, the IF results reported in this work are expressed in terms of $\tan\delta$ as a function of temperature. The tests were carried out using a DMA Q800 TA Instrument equipped with a liquid nitrogen cooling system. The samples were tested in the single cantilever configuration at 0.1, 1 and 10 Hz in the temperature range of -130 to 400 °C with a heating rate of 2 °C/min. The width (1 mm) and the thickness (1 mm) of the wire were measured by a micrometer (±0.01 mm). The length of the specimens was 17.5 mm.

2.3.9 Thermal Analysis

Thermal analysis of powders were performed using a Setaram Labsys TG-DSC-DTA 1600 instrument in a flowing argon atmosphere. Micro- and nano-sized powder samples were tested to investigate phase transformations upon heating and cooling. Continuous heating experiments were carried out from RT up to 600 °C at an heating rate of 30 °C/min. cooling was then imposed down to RT by the same rate of 30 °C/min.

References

1. M.A. Trunoc, M. Schoenitz, X. Zhu, E.L. Dreizin, Effect of polymorphic phase transformations in Al_2O_3 film on oxidation kinetics of aluminum powders. Combust. Flame **140**, 310–318 (2005)
2. X. Phung, J. Groza, E.A. Stach, L.N. Williams, S.B. Ritchey, Surface characterization of metal nanoparticles. Mater. Sci. Eng., A **359**, 261–268 (2003)
3. M. Balog, F. Simancik, M. Walcher, W. Rajner, C. Poletti, Extruded $Al–Al_2O_3$ composites formed in situ during consolidation of ultrafine Al powders: effect of the powder surface area. Mater. Sci. Eng. A **529**, 131–137 (2011)
4. K. Wafers, C. Misra, in *Oxides and hydroxides of aluminum*. Alcoa Technical Report No. 19 Revised, vol. 64. (Alcoa Laboratories, 1987)
5. B. Rufino, F. Boulc'h, M.V. Coulet, G. Lacroix, R. Denoyel. Influence of particles size on thermal properties of aluminium powder, Acta Materialia. **55**, 2815–2827 (2007)
6. M. Balog, P. Krizik, M. Nosko, Z. Hajovska, M.V. Castro Riglos, W. Rajner, D.S. Liu, F. Simancika. Forged HITEMAL: Al-based MMCs strengthen with nanometric thick Al_2O_3 skeleton, Mat. Sci. Eng. A. **613**, 82–90 (2014)
7. R. Casati et al. Microstructure and damping properties of ultra fine grained Al wires reinforced by Al_2O_3 nanoparticles. Light Metals 2014 Edited by: John Grandfield TMS (The Minerals, Metals & Materials Society) (2014)
8. G.K. Williamson, W.H. Hall, X-ray line broadening from filed aluminium and wolfram. Acta Metall. **1**, 22–31 (1953)
9. Y.V. Milman, B.A. Galanov, S.I. Chugunova, Plasticity characteristic obtained through hardness measurement. Acta Metall. Mater. **41**, 2523 (1993)
10. J.H. Ahn, E.C. Jeon, Y. Choi, Y.H. Lee, D. Kwon, Derivation of tensile flow properties of thin films using nanoindentation technique. Curr. Appl. Phys. **2**, 525 (2002)
11. ISO/TR 29381:2008, in *Metallic materials—measurement of mechanical properties by an instrumented indentation test—indentation tensile properties* (2008)
12. B. Taljat, G.M. Pharr, Development of pile-up during spherical indentation of elastic–plastic solids, Int. J. Solids Struct. **41**(14), 3891–3904 (2004)
13. M.S. Blanter, I.S. Golovin, H. Neuhauser, H.R. Sinning, *Internal friction in metallic materials* (Springer, Berlin, 2007)

Chapter 3
Consolidations of Al Powder and Dry Al$_2$O$_3$ Nanoparticles

Abstract Commercially pure micro-sized Al powder, either in as-received condition or after high-energy ball milling processing, was used to verify the capability of ECAP and hot extrusion (HE) to consolidate metal powder. A ball-to-powder weight ratio r = 5:1 was adopted for grinding the metal powder for 5 h using 10 % of ethanol as PCA. Consolidation of powder was performed by ECAP, by hot extrusion and by combining the two process. ECAP was also used for consolidating Al-Al$_2$O$_3$ composite powder mixed by ball milling. The samples produced by these PM routes are summarized hereunder: (1) Pure Al powders consolidated by ECAP (3 steps at 200 °C), (2) Pure Al powders consolidated by HE (300 °C), (3) Pure Al powders consolidated by ECAP (3 steps at 200 °C) and subsequently processed by HE (300 °C), (4) Composite powders (2 and 5 wt% Al$_2$O$_3$) consolidated by ECAP (3 steps at 200 °C). It is to remark that processing temperatures and number of consolidation steps by ECAP were set based on previous experience and on preliminary tests. Microstructural investigations showed an inadequate dispersion of alumina particles within the matrix and the presence of large alumina clusters. Further samples were then prepared raising the milling time up to 24 h without modifying the ball to powder weight ratio and using 1 % of ethanol as lubricant. By this procedure, the following samples were produced: pure Al powders consolidated by ECAP (3 steps at 300 °C), and composite powders (2 wt% Al$_2$O$_3$) consolidated by ECAP (3 steps at 300 °C).

Keywords Powder metallurgy · Metal matrix nanocomposites · Aluminum · Alumina nanoparticles · ECAP · Extrusion

3.1 Powder Characterization

SEM investigation was performed on pure aluminum powders at several stage of the ball milling in order to monitor the effect of the processing time on the evolution of powder morphology.

© The Author(s) 2016   47
R. Casati, *Aluminum Matrix Composites Reinforced with Alumina Nanoparticles*,
PoliMI SpringerBriefs, DOI 10.1007/978-3-319-27732-5_3

Fig. 3.1 SEM micrographs
of **a** dry Al$_2$O$_3$ nanoparticles
and **b** Al powder as-received
[2]

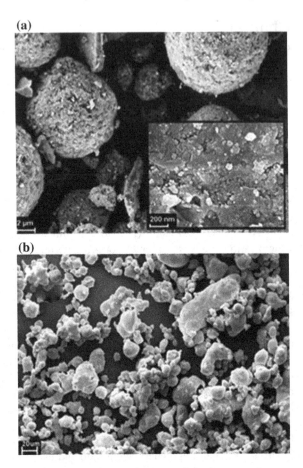

Fig. 3.1 SEM micrographs of **a** dry Al$_2$O$_3$ nanoparticles and **b** Al powder as-received [2]

In Fig. 3.1, the Al powder and the Al$_2$O$_3$ particles in the as-received condition are depicted. The aluminum powder particles look roughly spherical, smooth and discrete, while the alumina nanoparticles are agglomerated in micro-sized spherical clusters.

In Fig. 3.2, the view of the powder particles of the Al samples after 2 and 5 h of milling are shown. From quasi-spherical shape, the powder particles underwent morphological modifications promoted by the repetitive deformation, fracture and welding phenomena that occurred during ball milling [1].

After 5 h of milling, the Al particles showed a flake-like shape. High-energy ball milling process was carried out on Al-2 wt% Al$_2$O$_3$ powders with the aim of breaking the nanoparticle aggregates and dispersing the nano-reinforcement into the metal particles. Composite powder showed similar morphologies to the one depicted in Fig. 3.2 for pure Al. Alumina clusters were not noticeable at low magnifications.

Fig. 3.2 SEM micrographs of Al powder after **a** 2 and **b** 5 h of ball milling [2]

The morphology of pure Al and Al-2 wt% Al_2O_3 powders processed for 24 h are shown in Fig. 3.3. Both the powder samples looked as conglomerate of smaller particle fragments. In this case, the amount of lubricant used for the milling was lower than that used in the case of powder ground for 5 h, the powder were then more prone to weld together in cluster resulting in a less thin and flaky shape [1]. From the morphological point of view, the addition of 2 % of hard reinforcement did not significantly enhance the effect of the milling process.

It can be simply gather from the microstructures of Figs. 3.2 and 3.3 that the high-energy ball milling led to the plastic deformation of the metal powder particles. The accumulation of lattice defects was confirmed by XRD analysis. The full diffraction patterns of the pure Al powder before and after milling are shown in

Fig. 3.3 SEM micrograph of **a** pure Al and **b** Al-2 wt% Al$_2$O$_3$ powder after 24 h ball milling

Fig. 3.4, with a magnified view of the (111) peak at 38.4° in the inset. The milled powders exhibits broadened peaks, which indicate smaller crystallites and larger amount of crystal defects.

3.2 Powder Consolidation

Consolidation of the as-received and processed Al and composite powders was carried out via ECAP and hot extrusion. The powdered samples milled for 5 h were consolidated into Al alloy cans with thickness of 1 mm by ECAP at 200 °C without failures. This was not possible with the powder milled for 24 h. As previously mentioned in the introductory chapter, long milling times are responsible of high amount of residual strain into the metal matrix [1]. Therefore, powder ground for

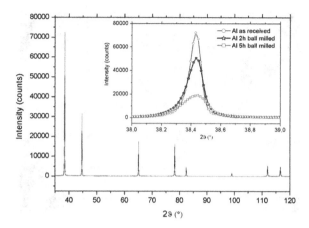

Fig. 3.4 XRD spectra of Al powder in as-received condition and after 2 and 5 h of high-energy ball milling. The magnification of the peak at 38.4° highlights the peak broadening with increasing milling time [2]

24 h exhibited lower capability to accommodate the necessary strain required by ECAP. Indeed, ECAP consolidation process mostly relies on severe plastic deformation of metal matrix particles [3]. For this reason, consolidation of powder can be achieved at such low temperature only when pressing rather soft powders. Therefore, the highly deformed powders that underwent to the high-energy grinding process for 24 h led to several cracks of the billet during processing at 200 °C (Fig. 3.5). To overcome this problem, thicker Al cans (thickness of 2.5 mm) and higher processing temperature (300 °C) were adopted. Nevertheless, even if by thicker cans and higher processing temperature the formation of cracks was fully avoid, lot of microstructural defects were observed in sintered samples. Pores and infiltration of the canning material were indeed well noticeable in the microstructure of the sample (Fig. 3.6).

Hot extrusion was carried out on Al powder milled for 5 h at 300 °C to promote formability of the Al powder and reduce the pressing loads. At this temperature, powder consolidation was regularly performed without any evidence of damage. Neither cracks nor morphological defects were observed in the extruded samples. Similarly, the same process and temperature (300 °C) were adopted to extrude an Al sample that was previously compacted by 2 ECAP passes at 200 °C. The extruded billet resulted free from cracks also in this case. This sample showed indeed good workability at high temperature.

Fig. 3.5 Al can containing Al powder milled for 24 h and processed by ECAP at 200 °C

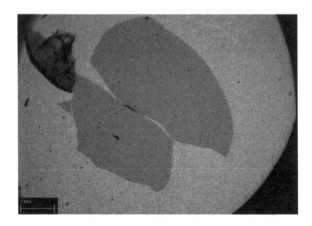

Fig. 3.6 SEM micrograph of 24 h milled powder processed by ECAP at 200 °C. The *brighter part* is the can material, while the *darker phase* is the consolidated Al powder. The can partially penetrated between the Al powder; pores (*black dots*) are also noticeable

3.3   Mechanical Characterization

The average microhardness values of the samples consolidated using both the as-received powder and the powder milled for 5 h are presented in Table 3.1. The samples ECAPed at 200 °C revealed to possess the higher hardness. HE and the combination of the two process (ECAP and HE) led to slightly lower hardness values. This might be due to several reasons, but the higher processing temperature (300 vs. 200 °C) is probably the main origin of the lower hardness numbers. Comparing the results of the as-received and the ball-milled samples consolidated

Table 3.1 Vickers microhardness of consolidated powder samples before and after high-energy ball milling for 5 h

Powder sample	Consolidation process	Microhardness (HV)	Density (g/cm³)
Al as-received	3 passes ECAP 200 °C	48.8	2.69
Al as-received	HE 300 °C	42.6	2.68
Al as-received	3 passes ECAP 200 °C + HE 300 °C	43.0	2.69
Al 5 h BM	3 passes ECAP 200 °C	51.5	2.67
Al 5 h BM	HE 300 °C	43.3	2.66
Al 5 h BM	3 passes ECAP 200 °C + HE 300 °C	43.6	2.68
Al/2 %Al₂O₃ 5 h BM	3 passes ECAP 200 °C	54.3	2.62
Al/5 %Al₂O₃ 5 h BM	3 passes ECAP 200 °C	58.2	2.53

via ECAP, an increase of 5.5 % of hardness Vickers was recorded as consequence of the work hardening induced in the metal powder through the grinding process.

The addition of 2 and 5 wt% of alumina had also a role in increasing the hardness of the samples. 5.4 and 13 % of increment was achieved by adding 2 and 5 % of reinforcement, respectively.

The results of density tests are also summarized in Table 3.1. The best results were shown by the sample compacted via ECAP, the theoretical density of pure Al (2.7 g/cm^3) was almost achieved. The powder consolidated after 5 h of milling showed lower density values than those consolidated starting from as-received powder. ECAP pressing seems to be able to better consolidate the Al particles than HE. For this reason, ECAP was chosen for consolidating the as received Al powder and the Al and composite powders milled for 24 h.

Much lower density values were reached by the composite samples reinforced with 2 and 5 % of alumina (2.62 and 2.53 g/cm^3, respectively) than for unreinforced Al powders.

In Table 3.2, the Vickers microhardness of the consolidated as-received powder samples and the 24-h milled powder samples are summarized. Considering the data of Tables 3.1 and 3.2, it can be noticed that when the temperature is increased from 200 to 300 °C, the samples produced using the as-received powder had a drop in hardness from 48.8 to 33.2 HV. It is reasonable to assume that the higher temperature was responsible for a more effective annealing action. Nevertheless, even if the sample produced using the powder subjected to 24 h milling was processed at higher temperature, it exhibited definitely higher microhardness values than the sample subjected to 5 h milling (86.3 vs. 51.5 HV). Thus, the higher temperature was not able to completely "erase" the effect of the longer milling time. The effect of 2 wt% of reinforcement was not noticeable, this sample showed indeed an hardness similar to that of the unreinforced Al that underwent the same process (86.4 HV).

3.4 Microstructural Characterization

In Fig. 3.7, a representative EBSD map of the CP Al sample milled for 5 h and consolidated by ECAP is depicted. This micrograph shows that the Al sample had a fine microstructure; the average grain size was estimated to be about 3 μm. The image quality (IQ) maps of Fig. 3.8 show that the boundaries of the Al powder

Table 3.2 Vickers microhardness of consolidated powder samples before and after high-energy ball milling for 24 h

Powder sample	Consolidation process	Microhardness (HV)
Al as-received	2 passes ECAP 300 °C	33.2
Al 24 h BM	2 passes ECAP 300 °C	86.3
Al/2 % Al$_2$O$_3$ 24 h BM	2 passes ECAP 300 °C	86.4

Fig. 3.7 Orientation map of
the Al sample

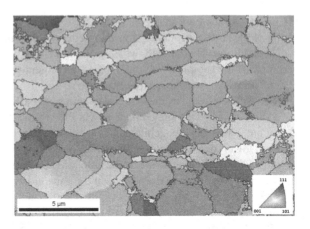

Fig. 3.8 IQ map of the Al
sample at **a** low and **b** high
magnification

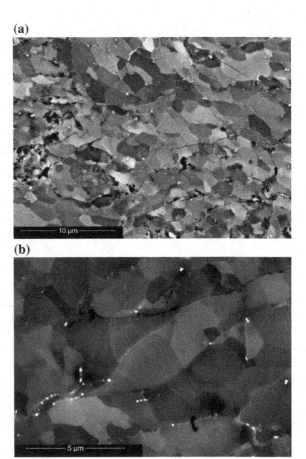

Fig. 3.9 Orientation map of the Al-2 wt% Al₂O₃ sample consolidated by ECAP

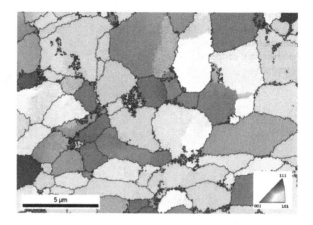

particles (black thin lines) are still visible in the consolidated sample. In Fig. 3.9, the EBSD map of the sample milled for 5 h and reinforced with 2 wt% of alumina is depicted.

This sample shows an average grain size comparable to the unreinforced sample. The ECAP revealed a suitable process to consolidate the Al powders without excessive grain growth. This is due to the low temperatures and times by which the process is carried out and to the high shear stress imposed to the green powders at the intersecting channel [3]. As expected, in both the above-mentioned specimens, no preferential grain orientation was detected. Microstructural investigations showed that the adopted PM route was not able to produce bulk nanocomposites featuring homogeneous dispersion of nanoparticles. Micrometric clusters of alumina (about 10–15 μm) were visible in samples that underwent 5 h BM and ECAP compaction (Fig. 3.10).

In some regions, evidence was found that the nano-alumina particles can also appear as well dispersed, as shown in Fig. 3.11. Figure 3.12 further shows that the alumina not only was found in form of big clusters and well dispersed nanoparticles, but also it was located in large amounts at the boundaries of the consolidated Al powder particles. This can hinder the consolidation process, leading to higher porosity, and justifies the lower density values of the composite samples shown in Table 3.1. Lower density values could also be due to the air entrapped in the big Al₂O₃ clusters shown in Fig. 3.10. Similar results were obtained with the sample reinforced with 5 wt% of alumina: some areas were characterized by well dispersed Al₂O₃ nanoparticles, but some clusters, in form of round particulate and spread at powder particle boundaries were noticed as well (Fig. 3.13).

Similar results were obtained with powders subjected to 24 h of ball milling. Higher milling time did not led to a significant improvement in dispersion of Al₂O₃ nanoparticles. Large clusters were well noticeable even by optical microscope (Fig. 3.14). Neither long milling time nor the high strain imposed by ECAP were able to break the alumina clusters.

Fig. 3.10 BSE micrograph of
Al-2 wt% Al$_2$O$_3$ and EDX
elemental analysis revealing
the chemistry of the Al$_2$O$_3$
aggregates

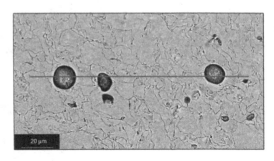

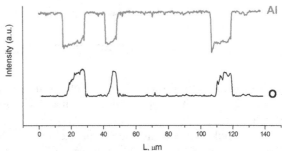

Fig. 3.11 BSE micrograph of
Al-2 wt% Al$_2$O$_3$ and EDX
elemental analysis

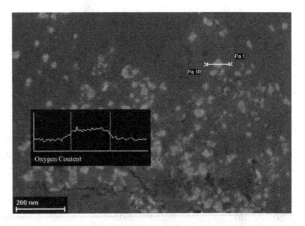

Fig. 3.12 SE micrograph of
Al-2 wt% Al$_2$O$_3$

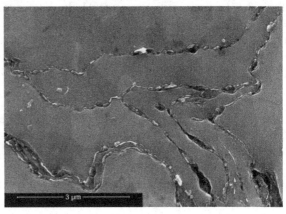

Fig. 3.13 Micrographs and EDX chemical analysis of the Al-5 wt% Al$_2$O$_3$ sample at **a** high and **b** low magnification, **c** micrograph showing the accumulation of alumina at the Al powder boundaries

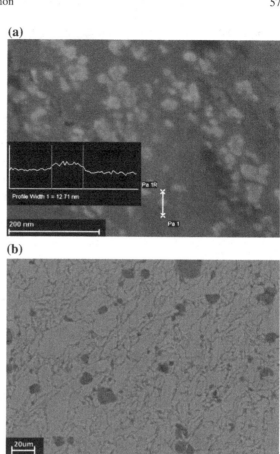

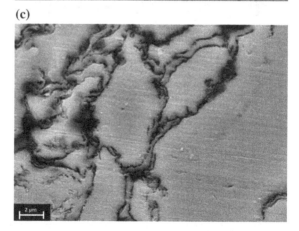

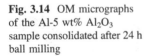

Fig. 3.14 OM micrographs
of the Al-5 wt% Al$_2$O$_3$
sample consolidated after 24 h
ball milling

References

1. C. Suryanarayana, Mechanical alloying and milling. Prog. Mater Sci. **46**, 1–184 (2001)
2. R. Casati, M. Vedani, D. Dellasega, P. Bassani, A. Tuissi. Consolidated Al/Al$_2$O$_3$ nano-composites by equal channel angular pressing and hot extrusion, Mat. Manuf. Process. **30**(10), 1–5 (2015)
3. X. Xia, Consolidation of particles by severe plastic deformation: mechanism and applications in processing bulk ultrafine and nanostructured alloys and composites. Adv. Eng. Mater. **12**, 724–729 (2010)

Chapter 4
Consolidation of Al Powder and Colloidal Suspension of Al$_2$O$_3$ Nanoparticles after 2 h Ball Milling

Abstract Inhomogeneous dispersion of alumina NPs was achieved by following the PM route described in the previous section. For these processing routes, dry Al$_2$O$_3$ nanoparticles and micro-sized Al powder were employed as starting materials. In particular, it was understood that high-energy ball milling was not able to completely break up the large alumina clusters. In the following experiments, an isopropanol colloidal solution of alumina nanoparticles was used instead of the dry alumina nanoparticles that were used for the previous tests. High-energy ball milling was performed to grind two different batches of powder: (1) Aluminum powder without ex situ reinforcement addition; (2) Aluminum powder + 2 wt% Al$_2$O$_3$ NPs (after manual stirring, the composite powder was dried by means of an electric stove at 50 °C to remove the isopropyl alcohol). Ball milling with a ball-to-powder ratio of 5:1 was performed for 2 h and ECAP was used to consolidate the pure and composite powders. 12 ECAP passes at 300 °C were then carried out to further break the potential Al$_2$O$_3$ clusters.

Keywords Powder metallurgy · Metal matrix nanocomposites · Aluminum · Alumina nanoparticles · ECAP · Extrusion · Ball milling

4.1 Powder Characterization

The employed Al powder were the same used for the previous experiments and depicted Fig. 3.1. The alumina morphology (after drying from the isopropanol solution) was investigated by TEM. Figure 4.1 shows that the oxide NPs had a roughly spherical structure, and their size was definitely lower than 50 nm. By indexing the corresponding selected area electron diffraction (SAED) ring-patterns, it can be stated that the ceramic reinforcement corresponded to the cubic γ-Al$_2$O$_3$ phase (JCPDS reference no. 10-0425). Al and Al$_2$O$_3$ particles were dried and mechanically milled for 2 h; their morphology is shown in Fig. 4.2. As previously described, the powdered samples looked as rough-edge conglomerate of fragment of particles.

© The Author(s) 2016   59
R. Casati, *Aluminum Matrix Composites Reinforced with Alumina Nanoparticles*,
PoliMI SpringerBriefs, DOI 10.1007/978-3-319-27732-5_4

Fig. 4.1 TEM micrograph of alumina nanoparticles after drying from isopropanol solution [1]

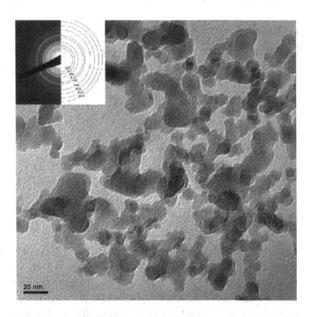

Fig. 4.2 SEM micrograph of the Al-2 wt% Al$_2$O$_3$ composite powder after 2 h milling

Also for this case it can be stated that from the morphological point of view, the addition of 2 % of hard reinforcement did not significantly enhance the effect of the milling process.

4.2   Powder Consolidation

Consolidation of pure and composite powdered samples were processed by ECAP at 300 °C using aluminum alloy cans (internal Ø 8 mm). 12 passes were performed following the so-called route *Bc* and no cracks occurred. The samples had a density

Table 4.1 Vickers microhardness of samples consolidated before and after high-energy ball milling for 2 h

Powder sample	Consolidation process	Microhardness (HV)	Density (g/cm^3)
Al 2 h BM	1 passes ECAP 300 °C	37.9	2.69
Al 2 h BM	12 passes ECAP 300 °C	38.0	2.69
Al/2 % Al$_2$O$_3$ 2 h BM	1 passes ECAP 300 °C	38.6	2.69
Al/2 % Al$_2$O$_3$ 2 h BM	12 passes ECAP 300 °C	44.7	2.69

of 2.69 g/cm^3 (Table 4.1). This value is close to the theoretical density of pure Al (2.70 g/cm^3); therefore, low porosity is expected.

4.3 Mechanical Characterization

Mechanical performance of consolidated samples was measured in term of microhardness. The samples showed quite low Vickers hardness. The milled Al powder samples consolidated either by 1 or 12 ECAP passes featured about 38 HV, while that of the composite sample was slightly higher: 38.6 HV after 1 ECAP pass and 44.7 HV after 12 ECAP passes. The increase in hardness might suggest a slightly improved dispersion of nanoparticles induced by the SPD process.

4.4 Microstructural Characterization

After powder consolidation performed by pressing the sample once through the ECAP channels, the original shape of the Al particle was still noticeable. They were marked by boundaries that were expected to be rich in oxygen due to the air exposure at 300 °C. The 12 ECAP passes were able to further fragment the oxide layer into smaller pieces and to disperse them in the Al matrix (Fig. 4.3).

These small oxide particles might have a slight strengthening effect, which was however not confirmed by hardness tests. The ECAP process was carried out for 12 passes at 300 °C, each sample spent about 20 min per pass into the hot die. This might have led to the annealing of the defects and to the release of the internal stresses accumulated during the ball milling process. This phenomenon might have erased the contribution in strength conferred by the oxide dispersion. Considering the microstructure of the composite sample of Fig. 4.4, it is possible to assert that the colloidal suspension of γ-Al$_2$O$_3$ did not lead to the formation of big clusters as in the case of the dry Al$_2$O used for the above-mentioned experiments (Figs. 3.10, 3.13 and 3.14).

Nonetheless, in the case of the composite sample after 1 ECAP pass (Fig. 4.4), the boundaries of the original Al particles were still well visible, while discrete

Fig. 4.3 SEM micrograph of the Al sample after **a** 1 ECAP pass and **b** 12 ECAP passes

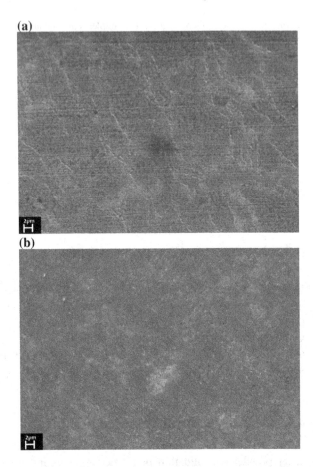

Fig. 4.4 SEM micrograph of the Al-2 wt% Al_2O_3 composite after 1 ECAP pass

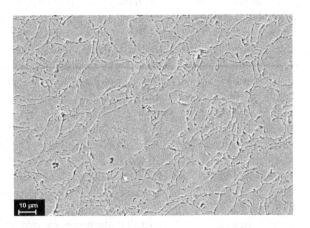

particles dispersed in the Al matrix were not noticed. Therefore, it can be supposed that the γ-Al$_2$O$_3$ nanoparticles were placed at the boundaries between the consolidated Al powder particles. After 12 ECAP passes, these boundaries rich in nanoparticles appeared to be broken into smaller particles. Nonetheless, the nanoparticles were not yet homogeneously dispersed throughout the matrix, but small clusters decorated the Al powder particle boundaries (Fig. 4.5). It is possible to conclude that the colloidal solution of Al$_2$O$_3$ nanoparticles did not lead to the formation of big reinforcement clusters as it was the case for the dry Al$_2$O$_3$ used in previous experiments.

Nevertheless, the ball milling process was not enough severe to fully embed the nanoparticles into the Al matrix. 12 ECAP passes were able to break the boundaries of the consolidated Al powder particles where the oxide nanoparticles were located, but they were not able to homogeneously disperse the nano-reinforcement into the Al matrix.

Fig. 4.5 SEM micrograph of the Al-2 wt% Al$_2$O$_3$ composite after 12 ECAP passes [2]

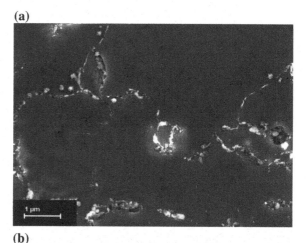

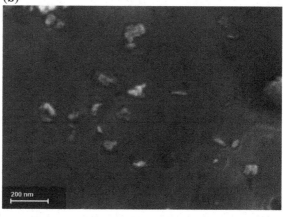

References

1. R. Casati, A. Fabrizi, G. Timelli, A. Tuissi, M. Vedani. Microstructural and mechanical properties of Al-based composites reinforced with in-situ and ex-situ Al$_2$O$_3$ nanoparticles, Adv. Eng. Mat. doi:10.1002/adem.201500297
2. R. Casati, M. Amadio, C.A. Biffi, D. Dellasega, A. Tuissi M. Vedani. Al/Al$_2$O$_3$ nano-composite produced by ECAP. Mat. Sci. Forum **762,** 457–464 (2013)

Chapter 5
Consolidation of Al Powder and Colloidal Suspension of Al_2O_3 Nanoparticles After 16 h Ball Milling

Abstract In order to have a more homogeneous particle dispersion, a much more severe ball milling process was carried out. A higher ball-to-powder ratio (r = 10:1) was adopted and the attritioning process was performed for 16 h. So far, better dispersion was achieve by using colloidal solution of γ-Al_2O_3 nanoparticles in isopropyl alcohol than by using dry-Al_2O_3. Thus, the composites were prepared again by adding 2 wt% of colloidal alumina to the Al powder. The micrographs of the starting materials correspond to those shown in Figs. 3.1 and 4.1. Either ECAP at 400 °C or hot extrusion at 400 °C were performed to consolidate the composite powder after milling. A selection of the extruded billets were then cold rolled down to a cross section of 1 mm^2 to verify the workability of the material. Pure Al powder was also consolidated and cold worked following the same procedure adopted for the nanocomposites so as to investigate the properties of a reference unreinforced sample.

Keywords Powder metallurgy · Metal matrix nanocomposites · Aluminum · Alumina nanoparticles · ECAP · Extrusion · Ball milling · In situ · Ex situ · Oxide dispersion

5.1 Powder Characterization

As already shown in Fig. 3.1, the SEM morphological analysis revealed that the as-received Al particles were almost spherical, with smooth surfaces, generally not aggregated in clusters.

After 16 h of ball milling with r = 10:1, the Al powder particles appeared as agglomerated, more flat and with sharper corners (Fig. 5.1). The morphology of milled Al-2 wt% Al_2O_3 composite powder looked very similar to that of CP Al powder.

As expected, the severe grinding process not only strongly affected the morphology of the Al powder, but also the microstructure (Fig. 5.2). The as-received material exhibited an ultrafine equiaxed grain structure. During ball milling, the repeated collisions between balls and powder particles led to the repeated fragmentation and welding of the Al particles, which in turn induced a further

© The Author(s) 2016
R. Casati, *Aluminum Matrix Composites Reinforced with Alumina Nanoparticles*,
PoliMI SpringerBriefs, DOI 10.1007/978-3-319-27732-5_5

Fig. 5.1 SEM micrographs of the Al powder undergone 16 h ball milling at **a** low and **b** high magnification [33]

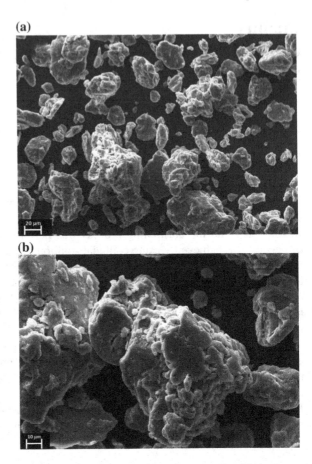

refinement of the grain structure. The grains turned out to be in the nanoscale range and more elongated, as depicted in Fig. 5.2b.

The XRD spectra of the as-received and ball-milled powders are depicted in Fig. 5.3. The severe ball milling led to a peak broadening and to the reduction of the peak intensities (see for instance the magnification of the peak at 38.4° referred to (111) planes in the inset of Fig. 5.3). Micro-strain and crystallite size of the powders were estimated from XRD results by using the Williamson-Hall method [1, 2].

The micro-strain of the as-received Al powder was evaluated to be 0.03 %, while after ball milling it revealed to be 0.17 % for both the CP Al and composite powders. The average crystallite size was measured to be about 660 nm for the as-received Al powder, and to be about 120 nm for the milled samples. The above-mentioned results are summarized in Table 5.1.

The mechanical response of the Al and composite powders was evaluated by Vickers hardness tests. The average hardness of the as-received Al powder was

Fig. 5.2 Microstructure of the Al powder. **a** Before and **b** after 16 h ball milling [34]

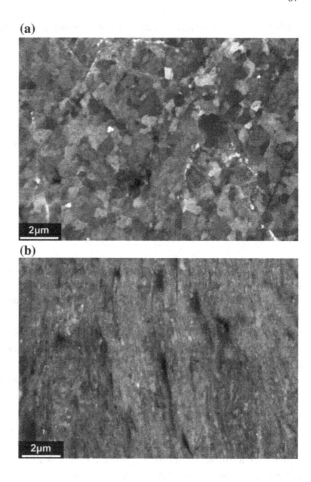

Fig. 5.3 XRD spectra of the Al powder in as-received condition and after 16 h ball milling and XRD spectrum of the Al-2 wt% Al$_2$O$_3$ composite powders after 16 h ball milling [34]

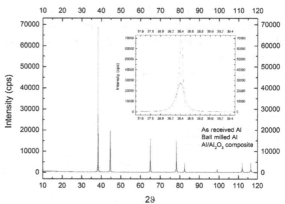

Table 5.1 Micro-strain and crystallite size of the powders estimated by Williamson-Hall method

Sample	Crystallite size (nm)	Micro-strain (%)
Al	662	0.03
Al 16 h BM	120	0.17
Al-2%Al$_2$O$_3$ 16 h BM	114	0.17

about 33 HV, while after severe ball milling process the hardness of the CP Al and of the Al-2 wt% Al$_2$O$_3$ composite powders increased up to about 84 and 98 HV, respectively.

5.2 Powder Consolidation

After ball milling, consolidations of pure and composite powdered samples were successfully performed without any occurrence of cracks neither by ECAP at 400 °C nor by extrusion at 400 °C. Cylindrical brass cans with external diameter of 10 mm and thickness of 1 mm were used as powder containers. Indeed, severe ball milling (16 h, r = 10:1) induced defects in the crystal lattice of powders and led to a raise of internal stresses, so that the metallic particles became less prone to be deformed and therefore compacted.

For this reason, in order to achieve a good compaction and to avoid cracks, consolidation processes had to be performed at a relatively high temperature. In the case of ECAP consolidation, 12 passes were performed following the route *Bc*. The samples were near-fully dense. The density of CP Al and composite after consolidation were 2.70 g/cm^3 and 2.71 g/cm^3, respectively. It is worth to mention that the density of alumina is about 3.9 g/cm^3, then a value higher than 2.70 g/cm^3, was expected for the composite sample. After powder consolidation via hot extrusion, the billets were cold rolled without failures down to square wires with cross section of 1 mm^2. The square section was constant along the length of the wires, which was about 300 mm. The average density of the wires was again 2.70 and 2.71 g/cm^3, for the CP Al and composite, respectively. A picture of the consolidated billets and wire is reported in Fig. 5.4.

Fig. 5.4 View of the ECAP and extrusion billets and of the rolled wire

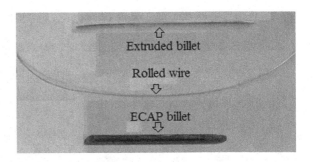

5.3 Mechanical Characterization of Samples Consolidated by ECAP

The results obtained by microhardness tests and depicted in Fig. 5.5 showed a reduction in hardness with increasing number of ECAP passes. The hardness decrease revealed by the CP Al specimen subjected to ball milling was significantly more drastic than that of the composite. After the first pass, the hardness of the pure Al and nano-composite was 107 HV and 117 HV, respectively; while, after 12 passes, it decreased to 76 HV and 111 HV, respectively.

Moreover, the mechanical behavior of the materials was measured by instrumented micro-indentation tests. Figure 5.6 shows the load versus indentation-depth curves when the material was cyclically stressed with stepwise increasing loads, as described in the previous chapter (Experimental methods).

These data were then converted according to Eqs. 2.2 and 2.3 (see Chap. 2) into true stress versus true strain charts given in Fig. 5.7, depicting the plastic behavior of the analyzed materials. It can be inferred that the Al-2 wt% Al_2O_3 nanocomposite showed higher strength than the CP Al material. After 12 ECAP passes, both the composite and the CP Al reduced their maximum strength.

These behaviors are consistent with the results of the microhardness tests. The different stress-strain behavior of the materials was also confirmed by their strain-hardening values, which were estimated to be 0.17 and 0.22 for CP Al and Al-2 wt% Al_2O_3 samples, respectively (Fig. 5.8).

Fig. 5.5 Evolution of microhardness for composite and pure Al samples during ECAP. High-energy ball milling is tagged in the legend as HEBM [35]

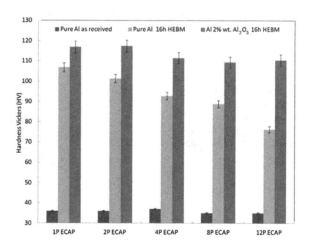

Fig. 5.6 Force-indentation
depth graphs showing the
behavior under mechanical
loading/unloading of the Al
and composite samples after
a 1 and **b** 12 ECAP passes
[34]

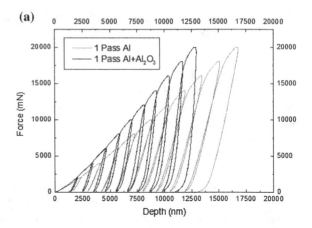

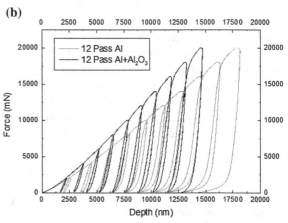

Fig. 5.7 Plastic flow curves
(true stress-strain) calculated
from the instrumented
indentation test results for the
Al and composite samples
after 1 and 12 ECAP passes
[34]

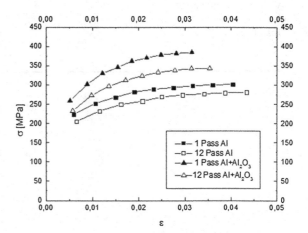

Fig. 5.8 Log-log stress-strain curves plotted to determine the strain-hardening exponents of the Al and composite samples after 1 and 12 ECAP passes [34]

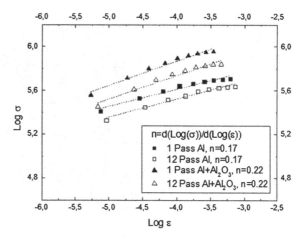

5.4 Microstructural Characterization of the Samples Consolidated by ECAP

The SEM microstructures of the powder samples processed by 1 and 12 ECAP passes are illustrated in Figs. 5.9 and 5.10, respectively. After 1 ECAP pass, the grain structure was not completely homogeneous, with evidence of relatively coarse grains (~ 1 μm in size) together with regions featuring very fine grains (even less the 100 nm in size). On the other hand, the microstructure after 12 passes became slightly coarser but more homogeneous. The addition of ceramic nanoparticles does not seem to induce any remarkable modification in the grain size of the material. By comparing the micrographs of the ball-milled powder (Fig. 5.2) with those of the consolidated materials, it is also highlighted that the size scale of grains increased owing to the consolidation process carried out at 400 °C.

The microstructures of transversal sections of pure Al and Al-2 wt% alumina specimens were also analyzed by means of EBSD technique. The corresponding orientation maps are depicted in Figs. 5.11 and 5.12. The EBSD analyses confirmed the slight grain growth occurring with increasing number of ECAP passes; the average grain size was indeed estimated to be about 0.6 and 1 μm for the 1 and 12 ECAP pass samples, respectively, irrespective from presence of reinforcement. From the grain boundary misorientation angle distributions of the investigated samples (Fig. 5.13), it results that the majority of grain boundaries are HAGBs, as the fraction of LAGBs was estimated to be lower than 6 %. The number of ECAP passes as well as the addition of the γ-Al$_2$O$_3$ NPs did not seem to induce any significant variation in the misorientation angle distribution. Moreover, it is worth to be mentioned that the specimens did not present any particular texture or preferential grain orientation.

Typical TEM micrographs of the CP Al and Al-2 wt% Al$_2$O$_3$ samples consolidated by 1 ECAP pass are depicted in Figs. 5.14 and 5.15. The microstructure

Fig. 5.9 Microstructures
of CP Al samples after
a 1 and **b** 12 ECAP
passes [34]

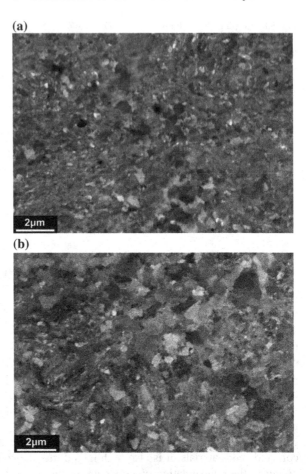

consisted of fine elongated grains and the corresponding SAED patterns showed
ring-patterns, indicative of polycrystalline materials with randomly oriented grains.
At higher magnification, the presence of nanoparticles was observed in the Al-2 wt%
Al_2O_3 composite both within the Al grains and at GBs. Some of these features turned
out to be discrete particles, while others reveal to be aggregated particles in the form
of small clusters. The grains of the CP Al also contained nanoparticles, with a lower
volume fraction when compared to Al-2 wt% Al_2O_3 composite. The origin of these
particles will be discussed in the next sections.

Results about the TEM analyses of the materials consolidated after 12 ECAP
passes are shown in Figs. 5.16 and 5.17. By comparing these microstructures with
those given in Figs. 5.14 and 5.15 related to 1 ECAP pass, it is possible to assert
that the 12 ECAP passes performed at 400 °C led to a coarser and near equiaxed
grain structure, as also shown by EBSD analysis. As already observed, also after 12
ECAP passes the CP Al showed nano-sized particles distributed throughout the
matrix.

Fig. 5.10 Microstructures
of composite samples after
a 1 and **b** 12 ECAP
passes [34]

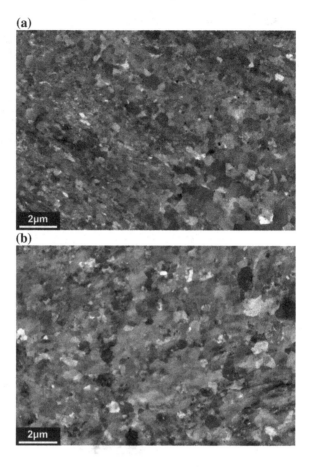

(a)

2μm

(b)

2μm

Higher amount of NPs was detected in the Al-2 wt% Al$_2$O$_3$ composite with a more homogeneous distribution than that reported for samples after 1 ECAP pass. However, few small NPs clusters were still noticeable in the microstructure.

The sample microstructures were also investigated by SEM after surface polishing and etching the by Keller's solution. The etchant had a strong effect in highlighting the nanoparticles. The metal matrix chemically reacted with the acid solution, while the oxide particles were inert to the etchant action. The drawback of this operation was the creation of small porosity probably due to the removal of particles. The microstructure of the samples pressed 12 times through the ECAP die are shown in Fig. 5.18 and they basically confirmed the TEM analysis. Very small and well dispersed nano-sized particles (<40 nm) were indeed found in the CP Al sample. The samples reinforced with 2 wt% of nano-sized Al$_2$O$_3$ exhibited higher amount of particles and the presence of some small clusters.

Fig. 5.11 Orientation maps
of CP Al sample after
a 1 and **b** 12 ECAP
passes [35]

(a)

(b)

5.5 Mechanical and Functional Characterization of Samples Consolidated by Hot Extrusion and Cold Rolled

Vickers hardness tests and tensile tests were carried out in order to estimate the mechanical properties of the wires produced through hot extrusion and cold rolling. The Al wires reinforced with the 2 wt% of alumina particles showed higher hardness than the CP Al wire (106 ± 1 HV vs. 96 ± 1 HV). These values are considerably higher than those typically found in CP Al (~ 20 HV) and in UFG CP Al (~ 40HV) [3]. The HV results are summarized in Table 5.2. In the same table, the results of tensile tests are reported as well, while typical tensile stress-strain curves are shown in Fig. 5.19. It is first noted that both tested wires (composite and unreinforced Al) showed YS and UTS values significantly higher than those generally found for CP Al. The wire reinforced with alumina nanoparticles added

Fig. 5.12 Orientation maps of Al-2 wt% Al₂O₃ sample after **a** 1 and **b** 12 ECAP passes [35]

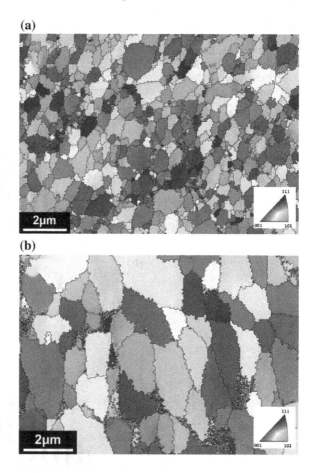

ex-situ showed both the highest yield strength (YS) and ultimate tensile strength (UTS). In particular, the material reinforced with 2 wt% of Al₂O₃ reached YS and UTS of 282 MPa and 373 MPa, while the wire produced from CP Al powder reached YS and UTS of 225 MPa and 302 MPa, respectively. By comparison, it has been reported that CP Al and UFG CP Al show much lower strength, i.e. YS of 20 MPa and 110 MPa and UTS of 30 MPa and 120 MPa, respectively [4].

The internal friction of the two wires was also evaluated by mechanical spectroscopy. DMA was employed for evaluating the tanδ of the materials as a function of temperature. In Fig. 5.20, the results of IF tests are depicted. At 0.1, 1 or 10 Hz, a tanδ peak is observed for both types the wires. The wire reinforced with ex-situ Al₂O₃ NPs showed improved damping performance at temperatures higher than 50 °C for 0.1 Hz, 70 °C for 1 Hz and 100 °C for 10 Hz. The tanδ value at 1 Hz for the monolithic CP aluminum is ∼0.001 at 25 °C and ∼0.007 at 275 °C [5]. Thus, a significant increase in IF was achieved in both the materials investigated. From

Fig. 5.13 Misorientation distributions of **a** CP Al and **b** Al-2 wt% Al$_2$O$_3$ samples consolidated by 1 and 12 ECAP passes; the black curves correspond to MacKenzie distribution [34]

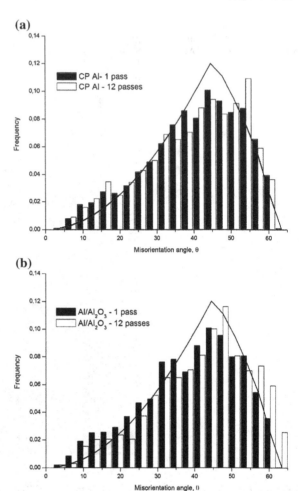

Fig. 5.20, the shift of the peaks towards higher temperatures can be clearly identified for both the materials when the loading frequency is increased.

The curves of Fig. 5.21 show that the storage modulus of the MMnC reinforced with ex-situ NPs is higher than that of the CP Al sample in the whole temperature range and for all of the frequencies investigated, although as temperature increases the difference in the storage modulus decreases. At room temperature the storage modulus values reached by the CP Al and the ex-situ reinforced MMnCs are about 64 GPa and 68 GPa, respectively.

Fig. 5.14 TEM micrographs taken at **a** low and **b** high magnification of CP Al sample consolidated by 1 ECAP pass [34]

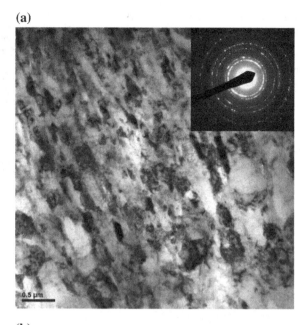

(a)

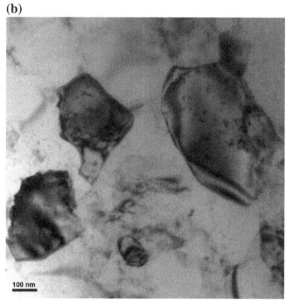

(b)

Fig. 5.15 TEM micrographs taken at **a** low and **b** high magnification of Al-2 wt% Al$_2$O$_3$ sample consolidated by 1 ECAP pass [34]

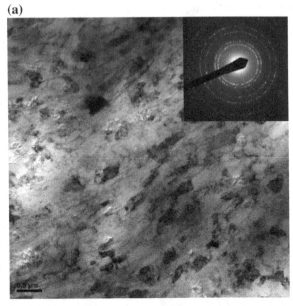

(a)

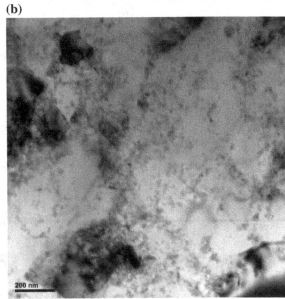

(b)

Fig. 5.16 TEM micrographs at **a** low and **b** high magnification of CP Al sample after 12 ECAP passes [34]

(a)

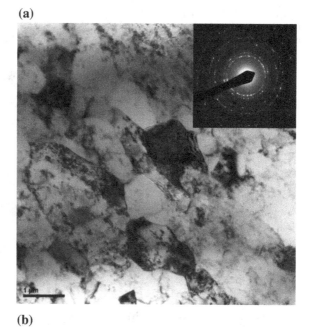

(b)

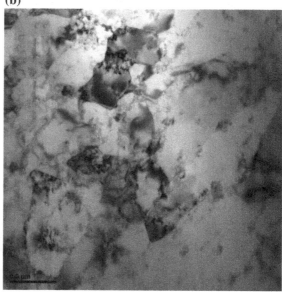

Fig. 5.17 TEM micrographs
taken at **a** low and **b** high
magnification of Al-2 wt%
Al_2O_3 sample consolidated by
12 ECAP passes [34]

(a)

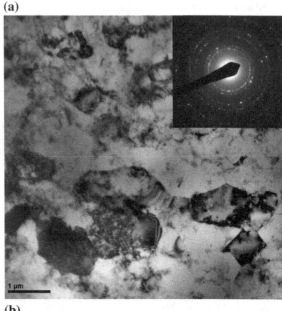

(b)

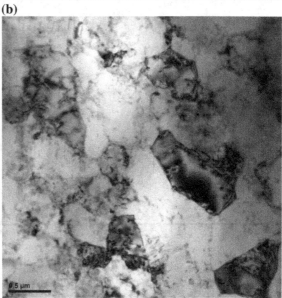

Fig. 5.18 SEM micrograph of the **a** CP Al and **b** Al-2 wt% Al$_2$O$_3$ samples consolidated by 12 ECAP passes. Samples were etched by Keller's solution [36]

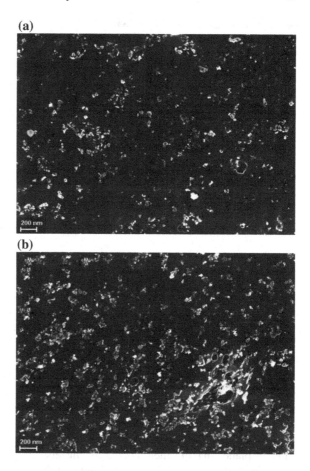

Table 5.2 YS, UTS and HV values of cold rolled wires

Samples	HV	YS (MPa)	UTS (MPa)
Al 16 BM	96 ± 1	225	302
Al-2%Al$_2$O$_3$ 16 h BM	106 ± 1	282	373
CP bulk Al	∼20 [3]	∼20 [4]	∼30 [4]
CP UFG bulk Al	∼40 [3]	∼110 [4]	∼120 [4]

The results are compared with those of CP bulk Al found in literature [3, 4]

5.6   Microstructural Characterization of Samples Consolidated by Hot Extrusion and Cold Rolled

The microstructure of the wires was investigated by SEM and TEM. In Fig. 5.22, the microstructure of the Al sample after etching is presented. As for the case of the samples consolidated by ECAP, the CP Al sample consolidated through extrusion

Fig. 5.19 Stress-strain curve
of Al and Al-2 wt%Al_2O_3
wires [33]

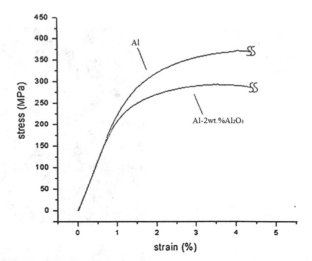

and then rolled showed particles distributed throughout the metal matrix almost free
of clusters. The particles were generally smaller than 40 nm. TEM-EDX chemical
mapping (Fig. 5.23) allowed to recognize those particles as composed by Al and O.
The particles are outlined by the oxygen-rich areas and their size was consistent
with that detected by SEM.

Figure 5.24 represents the microstructure of the composite wire reinforced with
2 wt% Al_2O_3 NPs. The fraction of particles is higher than that of CP Al and some
small clusters are present. It is worth to mention that the etching treatment is very
helpful for emphasizing the ceramic particles but, at once, it can also mislead the
analysis of the volume fraction of particles and it can emerge particles placed on
different planes making them appear as clusters.

In Fig. 5.25, the HAADF images highlight the grain structures in the two types
of wires, both showing ultrafine grains. The average grain size of the Al wire
reinforced with 2 wt% of ex-situ Al_2O_3 NPs was slightly smaller than that of the CP
Al wire (180 nm and 220 nm respectively, as measured by the linear intercept
method).

The crystallite size of the wires calculated by Williamson-Hall method showed
that during hot extrusion and repetitive annealing carried out during cold rolling,
the growth of the coherently scattering domains was limited (see Tables 5.1 and
5.3).

For the CP Al wire the crystallite size was estimated to be 228 nm (it was
120 nm before extrusion) and for the wire reinforced with the ex-situ particles, the
crystallite size was 206 nm (it was 114 nm before extrusion), in fairly good
agreement with the results from TEM. The results summarized in Table 5.3 also
show that the micro-strain decreased only slightly, even though the consolidation
and annealing were performed at 400 °C. It means that a high amount of lattice
defects is retained in the matrix.

Fig. 5.20 Tanδ versus
temperature curves at
a 0.1 Hz, **b** 1 Hz and
c 10 Hz [33]

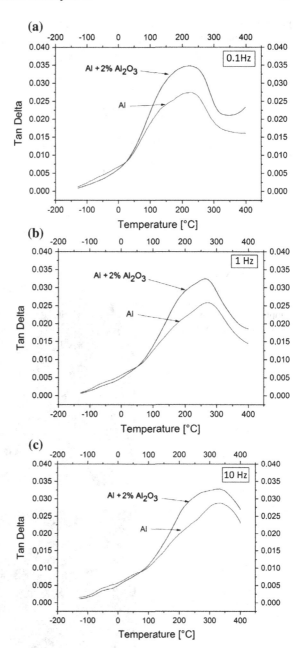

Fig. 5.21 Storage modulus
versus temperature curves
[33]

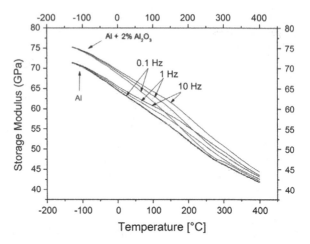

Fig. 5.22 SEM micrographs
of the CP Al at **a** low and
b high magnification. The
sample was etched by
Keller's solution [33]

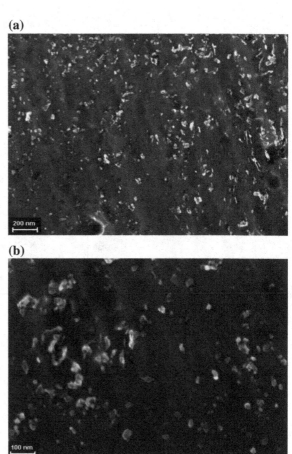

(a) **(b)**

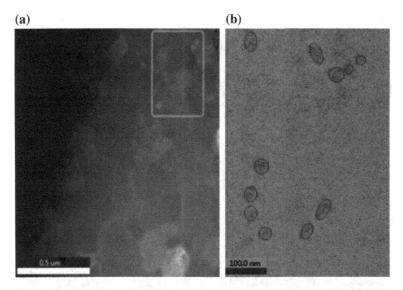

Fig. 5.23 a High-angle annular dark-field imaging (HAADF) image of the CP Al wire and **b** oxygen map of the area selected by the frame in **a** [33]

5.7 Discussion of the Results

Micrometric Al powder was firstly processed via high-energy ball milling for 16 h with a ball-to-powder ratio of 10:1. This process proved to be able to radically deform the metal particles in the plastic regime; their morphology changed because of the repeated welding, fracture and re-welding that occurred during milling (Figs. 5.1 and 5.2) [6]. Moreover, it has been reported [7–12] that the Al powder particles are covered by a very thin layer of oxide in the amorphous phase, whose thickness was estimated to be lower than 4 nm at room temperature. Therefore, it is reasonable to assume that also this layer was fragmented by the ball milling processing and that the resulting alumina debris were embedded into the metal matrix.

A second batch of Al powder was attritioned by planetary mill with a further addition of 2 wt% nano-sized γ-Al_2O_3, i.e. by ex-situ addition of the reinforcement. Before milling, the mix of nano γ-Al_2O_3 and metal powder was dried. During this operation, the NPs are believed to create clusters around the Al powders because of their high surface energy. Thus, high-energy ball milling was also aimed at breaking these clusters into discrete NPs and embedding them into the Al metal powder.

Accordingly, it is proposed that the grinding process may lead to two types of nanocomposite powder:

(i) the milled CP Al powder generates MMnCs powder reinforced with relatively small amount of alumina particles formed in-situ by fragmentation of the surface native oxide layers; and

(a)

(b)

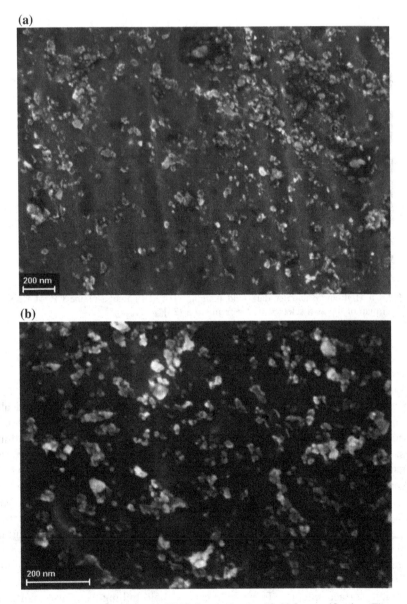

Fig. 5.24 SEM micrograph of the Al-2%Al$_2$O$_3$ at **a** low and **b** high magnification. The sample was etched by Keller's solution [33]

(ii) the Al powder milled with 2 wt% of γ-Al$_2$O$_3$ NPs generates MMnCs powder reinforced by a combination of alumina particles produced in-situ and alumina particles added ex-situ.

Fig. 5.25 HAADF images of **a** CP Al wire and **b** Al wire reinforced with 2 wt% ex-situ Al$_2$O$_3$ NPs [33]

(a)

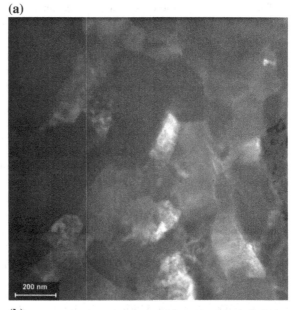

(b)

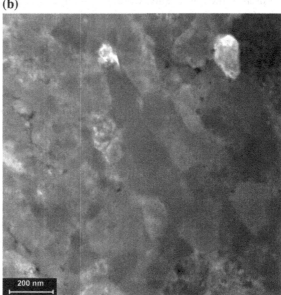

Table 5.3 Micro-strain and crystallite size of the powders estimated by Williamson-Hall method

Sample	Crystallite size (nm)	Micro-strain (%)
Al 16 h BM + HE + CR	228	0.10
Al-2%Al$_2$O$_3$ 16 h BM + HE + CR	206	0.12

These hypotheses were proven by SEM and TEM analyses carried out on the powder consolidated by ECAP (Figs. 5.14, 5.15, 5.16, 5.17 and 5.18) and by hot extrusion and then cold rolled (Figs. 5.22, 5.23 and 5.24). The in-situ particles appear to be nano-sized and homogenously distributed into the Al matrix (Figs. 5.14, 5.16, 5.18, 5.22 and 5.24). In the sample produced by ECAP, the in-situ NPs were observed by TEM both into the grains and at the grain boundaries. No information about the position of NPs in the microstructure is available for the samples produce by hot extrusion.

Moreover, by carefully observing the microstructure of the rolled CP Al sample (Fig. 5.22), some oxide fragments aligned along the rolling direction seem to show a size which is in agreement with the thickness of the native oxide layer measured in previous research works (<4 nm) [7–12].

Owing to their small sizes, these oxide NPs are expected to contribute to material strengthening, acting as barriers for dislocation slip, when they are not located at grain boundaries. Hampering the movement of dislocations and their bowing around NPs is considered by the Orowan strengthening effect [13, 14]. The contribution in strength associated to this phenomenon depends on the volume fraction and size of particles, as described in the introductory chapter.

The very small value of particle size is expected to significantly affect the final strength of the composite. Moreover, with the addition of extra (ex-situ) nano-sized γ-Al_2O_3 phase, the higher value of volume fraction is believed to further improve the strength of the composites. The microstructure of the Al-2 wt% Al_2O_3 nanocomposites confirmed that the oxide nanoparticles were successfully incorporated and distributed into the Al matrix both after the first ECAP pass (Figs. 5.15, 5.17 and 5.18) and after extrusion and rolling (Fig. 5.24), but few small NPs clusters remained in the material.

For a better understanding of the proposed mechanism, a schematic plot summarizing the effects of the whole process on the materials developed is depicted in Fig. 5.26.

The high-energy ball milling also induced a strong effect on the metal-matrix microstructure. The high amount of defects, likely in the form of dislocations, introduced by the repeated collisions of balls onto the powder particles, led to higher internal strain, evaluated to reach a value of 0.17 %. Moreover, the typical equiaxial grain structure of the as-received powder was radically changed. Smaller and more elongated grains were obtained and sub-micrometric porosity became noticeable (Fig. 5.2). Since GBs are considered as discontinuities in the crystal structure, they act as barriers to dislocation movement. This is due to the different orientations of adjacent grains and to the high lattice disorder characteristic of the boundary regions, which prevent the dislocations from moving in a continuous slip plane [15]. It is well known that the yield strength of a material varies with the grain size according to the Hall-Petch relation. In the present work, the results of the Williamson-Hall analysis on XRD spectra were consistent with SEM results, revealing a sharp reduction of crystallite size after ball milling, from about 0.7–0.1 μm. Previous works also showed that the grain size reduces drastically during mechanical milling and a nanostructure is generally formed due to this

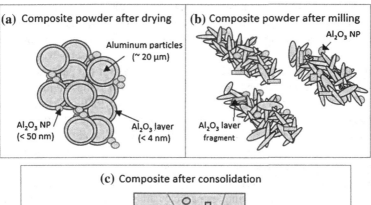

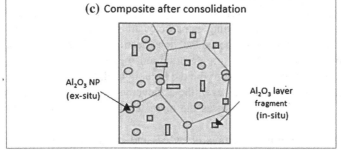

Fig. 5.26 Proposed mechanisms about the effects of milling and consolidation processes on the composite powder reinforced with in-situ and ex-situ NPs: **a** after drying, the aluminum powders, which are covered by an oxide passivation layer, are supposed to be additionally surrounded by γ-Al₂O₃ clusters; **b** after milling these clusters and the oxide passivation layer (*square fragments*) are broken up into small debris; **c** after ECAP compaction the fragments of passivation layer and the γ-Al₂O₃. NPs are dispersed into the aluminum matrix [33]

severe plastic deformation process [6]. These features, namely the reduction in the grain size, the higher amount of defects introduced by BM and the reinforcing effect given by the NPs dispersion, are supposed to induce a strong increase in material strength. Consistently, the hardness values of the Al particles increased by 154 % in the ball-milled powder reinforced with in-situ particles and by 195 % in the ball-milled powder reinforced with both in-situ and ex-situ particles.

The ECAP and the extrusion processes adopted in this work turned out to be suitable methods for the consolidation of Al micrometric powder. The formed billets were fully dense material substantially free of pores, as revealed by TEM investigations (Figs. 5.14, 5.15, 5.16, 5.17, 5.23 and 5.25). The nano-sized pores observed for instance in Fig. 5.25 are supposed to be artifacts, likely due to the detachment of particles due to the FIB cutting into slices less than 100 nm thick. The driving force for the consolidation in the conventional sintering processes is mainly due to the diffusion of atoms. In addition, bonding of particles via ECAP, which is based on SPD, also benefits from extensive plastic deformation of individual particles. The material plastic flow accompanying the deformation would fill any gap between particles, resulting in fully dense and well bonded bulk material [16]. For this reason, the consolidation could be achieved at temperatures (400 °C) lower than those used for conventional hot sintering for Al (about 600 °C) [17]. Kawasaki and co-workers, studying ECAP processing of a high purity bulk Al, reported an array of elongated cells or sub-grains after 1 ECAP pass [18]. In the present work, a similar microstructure was also obtained (Figs. 5.14 and 5.15) but the elongated grains are likely attributed not only to the ECAP processing but also to the original shape of the milled particles originated by the continuous ball-powder-ball collisions during the preliminary milling stages. Indeed, by accurately observing the micrographs of ball-milled powders (Fig. 5.2), fine elongated grains can be already perceived.

Similarly to the ECAP process, hot extrusion lead to the powder consolidation by deformation of metal particles and diffusion. Extrusion of powder is a continuous process mainly consisting in welding of particles, limited breaking of welds due to plastic flow, and re-welding as particles are rearranged and deformed in the extrusion die. It is actually supposed that densification might nearly be completed in the extrusion container during pre-deformation, before the actual extrusion takes place. Before the billet go through the die hole, indeed, the high pressure is able to compact the loose powder. However, the achievement of good structural integrity is dependent on shear deformation as the compacted powder is forced through the extrusion die [17].

Although both the temperatures of ECAP and extrusion/rolling processes were considerably high for pure Al (i.e. 400 °C), by comparing the microstructures of the BM powder (Fig. 5.2) with those of ECAP billets (Figs. 5.9, 5.10, 5.11, 5.12, 5.14, 5.15, 5.16 and 5.17) and especially with those of rolled wires (Figs. 5.4 and 5.6), it is possible to assert that grain growth was quite moderate. It can be reasonably assumed that grain growth was restrained by the high amount of oxide NPs that acted as pinning points on GB migration. Additionally, EBSD analysis carried out on ECAP samples showed that the orientation of grains seem to approach the condition of a complete random distribution, i.e. a MacKenzie distribution [19]. It is reported that from a certain strain level attained after extended milling times, sub-grains are formed due to the annihilation, recombination and rearrangement of the dislocations, and the sub-grain boundaries are transformed into randomly oriented HAGBs, due to reorientation of the single crystalline grains induced by SPD [20]. Even 12 ECAP passes did not impart any particular texture in both the nanocomposites samples.

The curves depicted in the σ-ε chart of Fig. 5.7 show the plastic flow of the CP Al and Al-2 wt% Al_2O_3 samples after 1 and 12 ECAP passes. The nanocomposites reinforced with ex-situ and in-situ nano-dispersoids showed higher strength than the material reinforced only with in-situ NPs (385 MPa and 302 MPa, respectively). This can be attributed to higher volume fraction of nanoparticles, which hinder the dislocation movements.

It is reported that UFG bulk Al samples processed by ECAP shows yield strength and ultimate tensile strength of 110 MPa and 120 MPa, respectively [4]. By comparing these values with the higher strength values achieved by the present

MMnCs, it is possible to assert that both in-situ and ex-situ nanoparticles have a remarkable effect on the mechanical strength of the materials.

However, it was established that after 12 ECAP passes the experimental materials showed a reduced strength compared to corresponding samples processed by only one single ECAP pass. Multi-passes ECAP was selected as an option to positively affect the reinforcement distribution throughout the material. However, maintaining the billets into the die at 400 °C led to a slight grain growth and probably to a reduction of the density of crystallographic defects (i.e. dislocations and GBs) that can also be responsible for the decreasing of the material strength.

The MMnCs produced by ECAP showed also different strain-hardening behavior: the strain-hardening exponent, n, was 0.17 for CP Al samples and 0.22 for the reinforced ones (Fig. 5.8). The increment in n can again be ascribed to the higher volume fraction of the nanoparticles. As aforementioned, the NPs have the capability to interact with the dislocations, increasing the strain-hardening [21].

The results of tensile tests carried out on the rolled wires, which are extrapolated from the curves of Fig. 5.9 and summarized in Table 5.2, show that the MMnC wires have a very high YS and UTS. Indeed, the wires showed strength values much higher than those generally related to pure CP bulk Al. In particular, the material reinforced with 2 wt% of Al_2O_3 reached YS and UTS of 282 MPa and 373 MPa, respectively while the wire produced from CP Al powder reached YS and UTS of 225 MPa and 302 MPa, respectively. By comparison, CP Al and UFG CP Al show much lower YS of 20 MPa and 110 MPa and UTS of 30 MPa and 120 MPa, respectively [4].

Again, the improvement is possibly due to the additive strengthening effects of dispersed nanoparticles and ultrafine grain size. The coefficient of thermal expansion (CTE) mismatch between the ceramic particles and the metal matrix should be also considered as potential strengthening mechanism [22–25]. This mismatch leads to the formation of dislocations, which are geometrically necessary to accommodate the different shrinking behaviors of the metal and the particles. Moreover, as for conventional metal matrix micro-composites, the load transfer from the soft matrix to the hard particles under an applied load contributes to the strengthening of the base material. The volume fraction and aspect ratio of nanoparticles in MMnCs are generally quite low, and therefore the contribution from load transfer is considered relatively small [26].

Internal friction tests were performed applying cyclic loads at frequencies of 0.1, 1 or 10 Hz. The tanδ vs. temperature curves of the MMnCs wire reinforced with in situ and in-situ/ex-situ NPs showed a peak at temperatures higher than 50 °C for 0.1 Hz, 70 °C for 1 Hz and 100 °C for 10 Hz (Fig. 5.20). The tanδ for the monolithic CP aluminum at 1 Hz was much lower ($\sim$0.001 at 25 °C and $\sim$0.007 at 275 °C) and it did not show any peak [5]. Moreover, from Fig. 5.20, the shift of peaks towards higher temperatures could be clearly identified for both materials when the loading frequency was increased.

It is known that the damping behavior is strongly sensitive to lattice defects (point defects, dislocations, grain boundaries and interfaces) [27]. The peak shift is generally related to relaxation processes (relaxation type peak) [27–29]. The higher density of defects (higher micro-strain) and higher amount of interfaces in the wire reinforced with ex-situ Al_2O_3 are thought to be the main responsible for the increased IF at high temperatures. In fact, although the external applied stress is small, the internal stress concentration in regions of high dislocation density may be large enough to cause relative atomic sliding. At room temperature, the displacements are typically fractions of an atomic diameter whereas, at high temperatures, sliding can be much more extensive and lead to viscoelastic strain. For this reason, the damping effects are more pronounced at elevated temperatures. However, with continued increase in temperature, dislocation mobility keeps increasing, resulting in annihilation of dislocations and eventually in a decrease in tan δ [27–29]. The higher modulus of the composites is believed to be due to the combination of the elastic moduli of alumina and aluminum.

The curves of Fig. 5.21 show the trend of the storage modulus when the temperature is changed in the range between −100 and 400 °C at 0.1, 1 and 10 Hz. At room temperature, the storage modulus reached by the in-situ and the ex-situ MMnCs are about 64 GPa and 68 GPa, respectively. The curves also show that the storage modulus of the MMnC reinforced with ex-situ NPs is higher than that of the CP Al sample in the whole temperature range and for all the frequencies investigated. This is a typical behavior of metallic materials and it is due the bonding force between atoms that decreases as the bond length increases, which happens during thermal expansion. Furthermore, as temperature increases, the difference in the storage moduli between the two wires decreases. This behavior can be ascribed to the relaxation phenomena as well. The microstructures undergoes annealing phenomena, able to reduce the density of crystal lattice defects.

5.8 Prediction of Experimental Results

The high strength of metal matrix nanocomposites is the result of several strengthening mechanisms that contribute to the final strength of the material. In particular, those for pure metals (without alloying elements) reinforced with nanoparticles are summarized hereunder:

(i)   load-transfer effect (or load-bearing effect) due to the transmission of load from the soft and compliant matrix to the stiff and hard particles [26];

(ii) Hall-Petch strengthening (or grain boundaries strengthening) which is related to the grain size of the metal matrix. The increase in volume fraction and the decrease of particle size lead to a finer structure, as theoretically modeled by the Zener equation [30, 31]

(iii) Orowan strengthening which is due to the capability of nanoparticles to obstacle the dislocation movement [13];

(iv) coefficient of thermal expansion (CTE) and elastic modulus (EM) mismatch, which are responsible of creating dislocation network around the particles [22–26, 32].

Moreover, work hardening contribution should be considered for those materials subjected to plastic deformation processes.

Some methods to estimate the final strength of materials were proposed in the open literature. Here, two methods are considered for calculation of the final strength considering the strengthening contributions above-listed:

(i) Method 1: the single strengthening mechanism contributions are directly summed to give the final strength of the composite. In this case, the super-position of the effects is not considered.

(ii) Method 2: the final strength of the composite is calculated as the mean square root of the single strengthening mechanism contributions. By this method, on the other hand, the superposition of the effects is considered.

More details and formulas of strengthening mechanism are described in the introductory chapter.

In Fig. 5.27, the curves of the theoretically calculated strengthening contributions and of the final strength of the nanocomposite as function of the particle size are shown. The strength values were calculated by considering a weight fraction of particles equal to 2.5 wt%. It is worth noting that the nanocomposite reinforced with ex-situ particles contained 2 % of γ-Al_2O_3 and about 0.5 % of native oxide. From the plot it is confirmed that all the strengthening mechanisms become weaker as the particle size increases. The CTE mismatch and the Orowan strengthening are the most effective mechanisms of strengthening when very small particles (5 nm) are used. For bigger particles, the CTE mismatch becomes the most effective contribution. The load transfer, as expected, has a minor role in conferring strength to the metal matrix since the volume fraction of NPs is very low. The values of the total increase in strength calculated by method 1 are unreliable for very small particle size values (above 1000 MPa). Vice versa, for bigger particle size, they become more reasonable and the gap between the total contribution in strength calculated by using Method 1 and Method 2 becomes smaller.

In Fig. 5.28 the curves of the single and total strengthening contributions of the nanocomposite as function of the weight fraction are shown. The strength values are calculated considering a particle size value equal to 20 nm. It is evident an

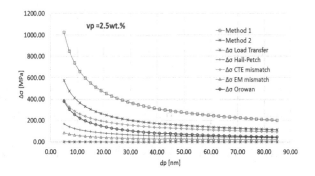

Fig. 5.27 Curves of the single and total strengthening contributions as function of the particle size, dp (weigth fraction = 2.5 wt%)

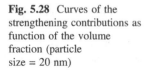

Fig. 5.28 Curves of the strengthening contributions as function of the volume fraction (particle size = 20 nm)

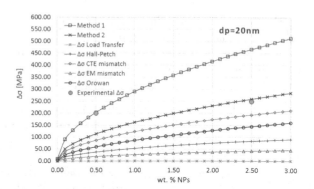

increase in single and total strengthening contribution when the volume fraction is increased.

The comparison of the contribution in strength values calculated using these literature formulas with the experimental data is not properly correct for several reasons: (i) the produced composites contained clusters, (ii) some particles were placed at the grain boundaries (then they are not efficient), (iii) the initial amount of native oxide is not precisely known, (iv) the in-situ produced NPs are not spherical but they have different aspect ratios, (v) the ex-situ added NPs size was not constant. Moreover, these models do not take into account the bonding force between NPs and metal matrix.

That being said, for sake of curiosity, the experimental results are compared with those predicted with the above mentioned models. Assuming that the NPs were round and their average size was 20 nm and the total weight fraction of NPs of the cold rolled in-situ reinforced composite wire was 0.5 %, then it is possible to notice that the experimental results in term of increase in strength well approach the value calculated with the Method 1. For the in-situ/ex-situ reinforce composite wire (weight fraction of NPs was about 2.5 %: 0.5 % in-situ NPs + 2% ex-situ NPs), the experimental value is quite far from that calculated with the Method 1, but it well approach the value calculated with the Method 2. Thus, it seems that the NPs are less effective, this could be due to the small clusters that were found in the microstructure.

References

1. A.K. Zak, W.H.A. Majid, M.E. Abrishami, R. Yousefi, X-ray analysis of ZnO nanoparticles by Williamson–Hall and size–strain plot methods. Solid State Sci. (2011)
2. G.K. Williamson, W.H. Hall, X-ray line broadening from filed aluminium and wolfram. Acta Metall. **1**, 22–31 (1953)
3. M. Saravanan, R.M. Pillai, B.C. Pai, M. Brahmakumar, K.R. Ravi, Equal channel angular pressing of pure aluminium—an analysis. Bull. Mater. Sci. **29**, 679–684 (2006)

4. V. Sklenicka, J. Dvorak, M. Svoboda, P. Kral, M. Kvapilova, Equal-channel angular pressing and creep in ultrafine-grained aluminium and its alloys, in *Aluminium Alloys—New Trends in Fabrication and Applications* (Chap. 1), ed. by Z. Ahmad. ISBN 978-953-51-0861-0, 5 Dec 2012 under CC BY 3.0 license

5. J.N. Wei, C.L. Gong, H.F. Cheng, Z.C. Zhou, Z.B. Li, J.P. Shui, F.S. Han, Low-frequency damping behavior of foamed commercially pure aluminum. Mater. Sci. Eng. A **332**, 375–381 (2002)

6. C. Suryanarayana, Mechanical alloying and milling. Prog. Mater Sci. **46**, 1–184 (2001)

7. M.A. Trunov, M. Schoenitz, X. Zhu, E.L. Dreizin, Effect of polymorphic phase transformations in Al_2O_3 film on oxidation kinetics of aluminum powders. Combust. Flame **140**, 310–318 (2005)

8. X. Phung, J. Groza, E.A. Stach, L.N. Williams, S.B. Ritchey, Surface characterization of metal nanoparticles. Mater. Sci. Eng. A **359**, 261–268 (2003)

9. M. Balog, F. Simancik, M. Walcher, W. Rajner, C. Poletti, Extruded $Al–Al_2O_3$ composites formed in situ during consolidation of ultrafine Al powders: effect of the powder surface area. Mater. Sci. Eng. A **529**, 131–137 (2011)

10. K. Wafers, C. Misra. Oxides and hydroxides of aluminum. Alcoa Technical Report No. 19 Revised, Alcoa Laboratories (1987), p. 64

11. B. Rufino, F. Boulc'h, M.V. Coulet, G. Lacroix, R. Denoyel, Influence of particles size on thermal properties of aluminium powder. Acta Materialia 55, 2815–2827 (2007)

12. M. Balog, P. Krizik, M. Nosko, Z. Hajovska, M.V. Castro Riglos, W. Rajner, D.S. Liu, F. Simancik. Forged HITEMAL: Al-based MMCs strengthen with nanometric thick Al_2O_3 skeleton. Mater. Sci. Eng. A. **613**, 82–90 (2014)

13. D. Hull, D.J. Bacon, *Introduction to Dislocations*, 4th edn. (Butterworth-Heinemann, London 2001)

14. Z. Zhang, D.L. Chen, Contribution of Orowan strengthening effect in particulate-reinforced metal matrix nanocomposites. Mater. Sci. Eng. A **483–484**, 148–152 (2008)

15. A. Sanaty-Zadeh, Comparison between current models for the strength of particulate-reinforced metal matrix nanocomposites with emphasis on consideration of Hall-Petch effect. Mater. Sci. Eng. A **531**, 112–118 (2012)

16. X. Xia, Consolidation of particles by severe plastic deformation: mechanism and applications in processing bulk ultrafine and nanostructured alloys and composites. Adv. Eng. Mater. **12**, 724–729 (2010)

17. Powder Metal Technologies and Applications was published in 1998 as Volume 7 of ASM Handbook

18. M. Kawasaki, Z. Horita, T.G. Langdon, Microstructural evolution in high purity aluminum processed by ECAP. Mater. Sci. Eng. A **524**, 143–150 (2009)

19. J. Mason, C. Schuh, The generalized Mackenzie distribution: Disorientation angle distributions for arbitrary textures. Acta Mater. **57**, 4186–4197 (2009)

20. D.B. Witkin, E.J. Lavernia, Synthesis and mechanical behavior of nanostructured materials via cryomilling. Prog. Mater Sci. **51**, 1–60 (2006)

21. S. Goussous, W. Xu, K. Xia, Developing aluminum nanocomposites via severe plastic deformation. J. Phys: Conf. Ser. **240**, 012106 (2010)

22. A. Sanaty-Zadeh, Comparison between current models for the strength of particulate-reinforced metal matrix nanocomposites with emphasis on consideration of Hall-Petch effect. Mater. Sci. Eng. A **531**, 112–118 (2012)

23. R. Casati, M. Vedani, Metal matrix composites reinforced by nano-particles—a review. Metals **4**, 65–83 (2014). doi:10.3390/met4010065

24. Z. Zhang, D.L. Chen, Consideration of Orowan strengthening effect in particulate-reinforced metal matrix nanocomposites: a model for predicting their yield strength. Scripta Mater. **54**, 1321–1326 (2006)

25. R.J. Arsenault, N. Shi, Dislocation generation due to differences between the coefficients of thermal-expansion. Mater. Sci. Eng. A **81**, 175–187 (1986)

26. V.C. Nardone, K.M. Prewo, On the strength of discontinuous silicon carbide reinforced aluminum composites. Scr. Metall. **20**, 43–48 (1986)
27. M.S. Blanter, I.S. Golovin, H. Neuhauser, H.R. Sinning, *Internal friction in metallic materials* (Springer, Berlin, 2007)
28. E. Carreño-Morelli, S.E. Urreta, R. Schaller, Mechanical spectroscopy of thermal stress relaxation at metal–ceramic interfaces in Aluminium-based composites. Acta Mater. **48**, 4725–4733 (2000)
29. J. Zhang, R.J. Perez, C.R. Wong, E.J. Lavernia, Effect of secondary phases on the damping behavior of metals, alloys and metal matrix composites. Mater. Sci. Eng.: R. **13**, 325–390 (1994)
30. E.O. Hall, The deformation and aging of mild steel. Proc. Phys. Soc. London, Sect. B **64**, 747–753 (1951)
31. N.J. Petch, The cleavage strength of polycrystals. J. Iron Steel Res. **174**, 25–28 (1953)
32. D. Hull, T.W. Clyne, *An Introduction to Composite Materials*, 2nd edn. Cambridge Solid State Science Series (1996)
33. R. Casati, X. Wei, K. Xia, D. Dellasega, A. Tuissi, E. Villa, M. Vedani, Mechanical and functional properties of ultrafine grained Al wires reinforced by nano-Al_2O_3 particles. Mater. Des. **64**, 102–109 (2014)
34. R. Casati, A. Fabrizi, G. Timelli, A. Tuissi, M. Vedani, Microstructural and mechanical properties of Al-based Composites reinforced with in-situ and ex-situ Al_2O_3 nanoparticles. Adv. Eng. Mater. doi:10.1002/adem.201500297
35. R. Casati, F. Bonollo, D. Dellasega, A. Fabrizi, G. Timelli, A. Tuissi, M. Vedani, Ex situ Al–Al_2O_3ultrafine grained nanocomposites produced via powder metallurgy. J. Alloy. Compd. **615**, S386–S388 (2014)
36. R. Casati, M. Amadio, C. Alberto Biffi, D. Dellasega, A. Tuissi, M. Vedani, Al/Al_2O_3 nano-composite produced by ECAP. Mater. Sci. Forum **762**, 457–464 (2013)

Chapter 6
Consolidation of AL Powder and Colloidal Suspension of Al₂O₃ Nanoparticles After 24 h Ball Milling

Abstract Additional experiments were performed with Al powder and alumina suspension adopting an increased milling time (24 h) to further improve the particles dispersion. The other milling parameters were the same used for the powder sample subjected to 16 h of ball milling. In particular, the same ball-to-powder ratio was used (r = 10:1). The powder were consolidated by a single ECAP pass at 400 °C. Samples reinforced with 0, 2 and 5 % of alumina were produced for comparison with the samples showed in the previous sections. Moreover, Al billets reinforced much higher NP fractions of 10, 20 and 30 wt% were prepared.

Keywords Powder metallurgy · Metal matrix nanocomposites · Aluminum · Alumina nanoparticles · Ball milling · Oxide dispersion

6.1 Powder Characterization

Although the different mix of powders (Al with 0, 2, 5, 10, 20 and 30 % nano-sized Al₂O₃) were ground using the same milling parameters, they revealed very different morphology (Figs. 6.1a, 6.2a, 6.3a, 6.4a, 6.5a and 6.6a).

As the amount of alumina was increased, the agglomeration of aluminum particles became increasingly difficult and it led to a reduction of the metallic particles size. The nano-sized ceramic compound were likely to work as abrasive agent leading to a more effective milling process. The high-magnification micrographs (100.000X) of the Al particles surface enables the observation of the alumina NPs (Figs. 6.1b, 6.2b, 6.3b, 6.4b, 6.5b and 6.6b). They appear as white spots and they look homogeneously dispersed on the Al particle surface, big alumina clusters are not noticeable.

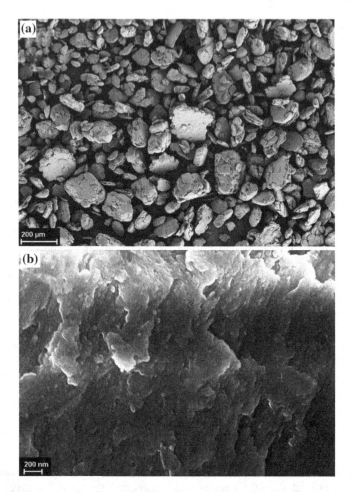

Fig. 6.1 CP Al powder after 24 h ball milling at **a** low and **b** high magnification

6.2   Powder Consolidation

Consolidations of pure and composite powdered samples were performed without any evidence of crack development when pressing the samples once through the ECAP channels at 400 °C. Cylindrical brass cans with increased thickness of 5 mm were employed to avoid failures during the process. As previously shown, the severe ball milling induces defects in the crystal lattice of the metal powders and

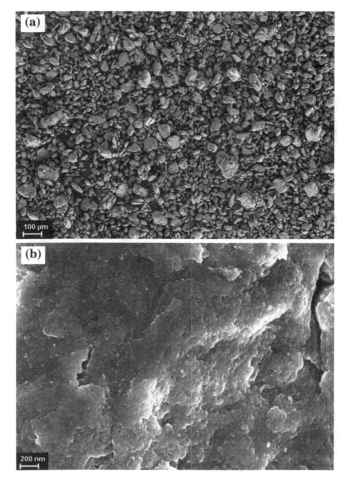

Fig. 6.2 Al-2 wt% Al$_2$O$_3$ composite powder after 24 h ball milling at **a** low and **b** high magnification

lead to a raise of internal stresses, thus the Al particles became less prone to be deformed and compacted at low temperatures. For this reason, in order to achieve a good compaction and to avoid cracks, consolidation processes had to be performed at relatively high temperature (400 °C) and thickness of cans had to be increased. Indeed, previous consolidation trials of milled Al powder were not successful at 300 °C. Moreover, even at 400 °C, consolidation of the most critic Al-30 wt% Al$_2$O$_3$ composite powder was just partially completed because of the very large amount of reinforcement.

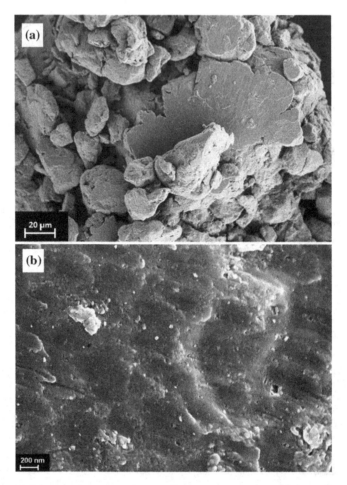

Fig. 6.3 Al-5 wt% Al_2O_3 composite powder after 24 h ball milling at **a** low and **b** high magnification

6.3 Mechanical Characterization

The average Vickers hardness obtained for the 24 h milled powder sample consolidated by 1 ECAP pass at 400 °C are summarized in Table 6.1 and in the histogram of Fig. 6.7.

For comparison, the results of the powder subjected to the same ball milling procedure but for a shorter time (16 h) and consolidated by 1 ECAP pass at the same temperature are reported in the graph. These latter were already discussed and reported in Table 5.2 (see also Ref. [1]).

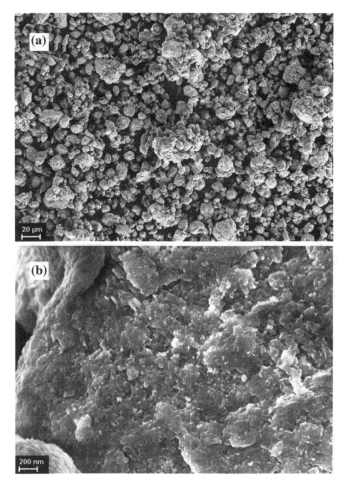

Fig. 6.4 Al-10 wt% Al$_2$O$_3$ composite powder after 24 h ball milling at **a** low and **b** high magnification

The histogram clearly show an almost linear increase in microhardness due to the increase in reinforcement fraction. This behavior is due to the different contribution in strength previously described. The only sample deviating from the above-mentioned trend was the composite reinforced with 30 % alumina owing to lack of full consolidation due to large amount of ceramic NPs. The longer ball milling, which is able to reduce the crystallite size and to increase the amount of internal stress as

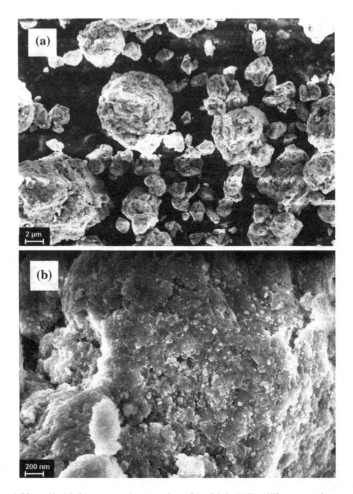

Fig. 6.5 Al-20 wt% Al_2O_3 composite powder after 24 h ball milling at **a** low and **b** high magnification

proved in the previous sections, was able to produce a further increase in hardness even if the process was carried out at relative high temperature (400 °C).

6.4 Microstructural Characterization

Microstructural characterization was carried out for the samples reinforced with 2 and 5 wt% of nanoparticles. The results are depicted in Fig. 6.8. The two samples showed a homogeneous distribution of particles. In this sense, 24 h ball milling led to slightly improved results over those achieved with less severe milling process.

Fig. 6.6 Al-30 wt% Al$_2$O$_3$ composite powder after 24 h ball milling at **a** low and **b** high magnification

Table 6.1 Hardness Vickers numbers of samples consolidated by 1 ECAP pass at 400 °C after 24 h ball milling

Powder sample	Consolidation process (°C)	Microhardness (HV)
Al as-received	1 ECAP pass 400	36
Al 24 h BM	1 ECAP pass 400	115
Al-2 % Al$_2$O$_3$ 24 h BM	1 ECAP pass 400	129
Al-5 % Al$_2$O$_3$ 24 h BM	1 ECAP pass 400	140
Al-10 % Al$_2$O$_3$ 24 h BM	1 ECAP pass 400	153
Al-20 % Al$_2$O$_3$ 24 h BM	1 ECAP pass 400	199
Al-30 % Al$_2$O$_3$ 24 h BM	1 ECAP pass 400	190

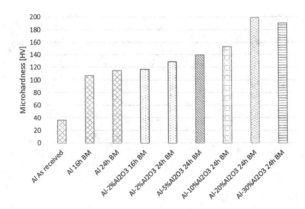

Fig. 6.7 Vickers hardness of samples consolidated by 1 ECAP pass at 400 °C after 16 and 24 h ball milling

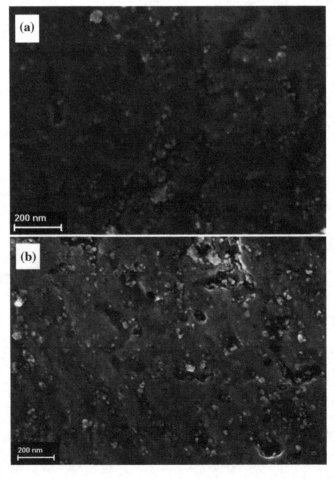

Fig. 6.8 SEM micrographs of the **a** Al-2 wt% Al_2O_3 and **b** Al-5 wt% Al_2O_3 composite samples

Reference

1. R. Casati, F. Bonollo, D. Dellasega, A. Fabrizi, G. Timelli, A. Tuissi, M. Vedani, J. Alloy. Compd. **615**, 386 (2014)

Chapter 7
Consolidation of Micro- and Nano-Sized Al Powder

Abstract In this final set of experiments, Al NPs were employed in the as-received and ball-milled condition to produce in situ reinforced MMnCs. The NPs possess higher surface than the micro-sized counterpart. This means that they may lead to the production of nanocomposites reinforced with a much higher content of in situ reinforcement and, as a limiting and desired condition, highly reinforced composites could be produced even without relying on ex situ addition of oxide NPs. For comparison, Al micro-sized powder was consolidated in the as-received condition and after ball milling as well. Furthermore, a mix of the two above-mentioned powders was also employed to complete the frame of experimental conditions. A ball-to-powder weight ratio r = 10:1 was adopted for grinding the metal powder for 16 h using 1.5 % of stearic acid as PCA. Powder consolidation was performed by BP-ECAP. It was expected that the higher content of non-metallic compound made the consolidation of powder rather difficult. It was also known that in SPD processes, more ductile and bigger particles ease the consolidation process [3] since the driving force is the severe plastic deformation of metal powder particles. On the contrary, nano-sized particles cannot accommodate high shear strains and are inclined to slip on each other instead of being deformed. Since the back-pressure (BP) revealed to be able to more efficiently consolidate powders by ECAP, it was applied for producing the nanocomposite billets. After preliminary attempts at different temperatures, 600 °C was selected as a suitable temperature for producing the following full dense bulk samples: (1) As-received Al micro-powders consolidated by BP-ECAP, (2) As-received Al nano-powders consolidated by BP-ECAP, (3) Ball-milled Al micro-powders consolidated by BP-ECAP, (4) Ball-milled Al nano-powders consolidated by BP-ECAP, (5) Ball-milled Al micro-(50 wt%) and nano-powders (50 wt%) consolidated by BP-ECAP.

Keywords Aluminum · Oxide dispersion · Nanoparticle · BP-ECAP · Phase transformation · XRD · DSC

© The Author(s) 2016
R. Casati, *Aluminum Matrix Composites Reinforced with Alumina Nanoparticles*,
PoliMI SpringerBriefs, DOI 10.1007/978-3-319-27732-5_7

107

7.1 Powder Characterization

Al powder (supplied by ECKA Granules Australia) was characterized by SEM analysis before and after 16 h ball milling. In Fig. 7.1, the starting micro- and nano-sized metal powder are shown. The micro-Al particles were quite rounded and discrete, showing an average size of about 20 μm.

The Al nanoparticles, were about 80 nm in diameter and turned out to be almost perfectly spherical. They seemed more agglomerated, but the individual particles were well visible in the SEM images. Their shape is consistent with considerations raised by Ramaswamy et al. [1] who stated that Al particles of 10–20 nm are crystallographic in shape, while they become spherical as the particle size increases.

As expected, after grinding the powder showed a completely different morphology (Fig. 7.2). The micro-sized powder particles had a flake morphology, while the nanoparticles mostly lose their nano-size and became aggregated into micro-sized porous clusters. Nonetheless, few discrete spherical nano-particles were still noticeable.

Fig. 7.1 As-received **a** micro-sized and **b** nano-sized Al powder [38]

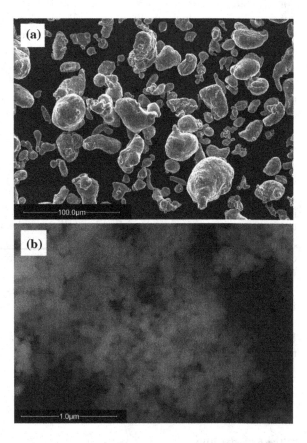

Fig. 7.2 Ball-milled
a micro-sized and
b nano-sized Al powder [38]

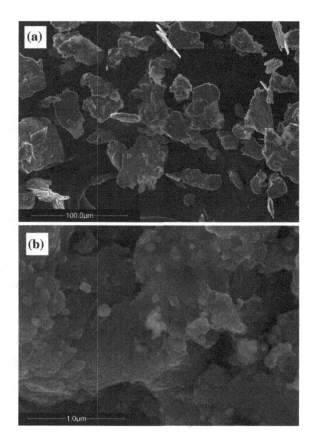

In Fig. 7.3, the SEM micrographs of the 50 % nano- 50 % micro-sized powder after ball milling are reported. At low magnification, they exhibited a flake-like shape, and at high magnification, they showed a surface morphology that was quite similar to that of the ball-milled nano-sized Al powder particles (Fig. 7.2).

7.2   Powder Consolidation

As previously mentioned, consolidation of powder was carried out by means of 4 ECAP passes applying a BP. Powders were wrap in a steel foil and cold pressed before consolidation. As shown in Table 7.1, the BP and the processing temperature were gradually increased until no cracks were noticeable on the consolidated billets. Whole billets free of cracks were therefore produced by adopting a BP of 220 MPa and a temperature of 600 °C.

Fig. 7.3 Ball-milled Al
micro-powders (50 %) and
nano-powders (50 %) at **a** low
and **b** high magnification [38]

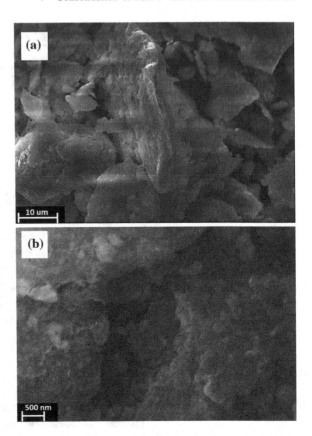

Table 7.1 Process parameters adopted for consolidation of Al powder via BP-ECAP

Powder sample	Consolidation process	Visual analysis
Nano Al as-received	4 ECAP pass 400 °C 50 MPa	Cracked
Nano Al as-received	4 ECAP pass 400 °C 180 MPa	Cracked
Nano Al as-received	4 ECAP pass 430 °C 220 MPa	Cracked
Nano Al as-received	4 ECAP pass 450 °C 220 MPa	Cracked
Nano Al as-received	4 ECAP pass 500 °C 220 MPa	Cracked
Nano Al as-received	4 ECAP pass 550 °C 220 MPa	Cracked
Nano Al as-received	4 ECAP pass 600 °C 220 MPa	Not cracked
Nano Al ball-milled	4 ECAP pass 600 °C 220 MPa	Not cracked
Micro Al as-received	4 ECAP pass 600 °C 220 MPa	Not cracked
Micro Al ball-milled	4 ECAP pass 600 °C 220 MPa	Not cracked
50 %Micro- 50 %Nano-Al ball-milled	4 ECAP pass 600 °C 220 MPa	Not cracked

In the third column, the visual aspect of the billet is described

7.3 Mechanical Characterization

Microhardness tests were performed on the consolidated samples and summarized in the histogram of Fig. 7.4.

The billets consolidated using the nano-sized powder showed extremely high hardness values (about 166 HV). It is worth mentioning that the Vickers hardness of pure bulk Al is about 20 HV [2], the in situ reinforcement therefore led to an hardness increase of about 800 %, while ball-milling did not lead to any significant hardness increment.

The samples produced by using the micro-sized Al powder showed much lower hardness values. Samples produced starting from the as-received powder showed a hardness of about 29 HV, while those prepared using the ball-milled powder exhibited a hardness of 62 HV. These values are consistent with the results previously achieved and they are the consequences of the in situ reinforcement and of the strain hardening produced during high-energy ball milling.

The as-received and ball-milled Al powder consolidated at 400 °C showed higher Vickers hardness (37 and 93 HV, respectively) due to a softer annealing action exerted during holding at ECAP temperature (400 vs. 600 °C). Finally, the sample produced starting from a mix of nano- and micro-particles showed a hardness of about 120 HV, which is an average value between the hardness of the sample produced using the ball-milled micro-sized Al powder and the ball-milled nano-sized Al powder.

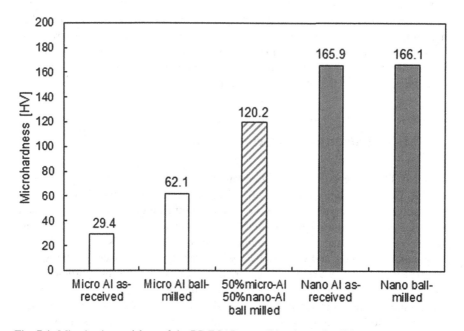

Fig. 7.4 Microhardness vickers of the BP-ECAP consolidated samples [38]

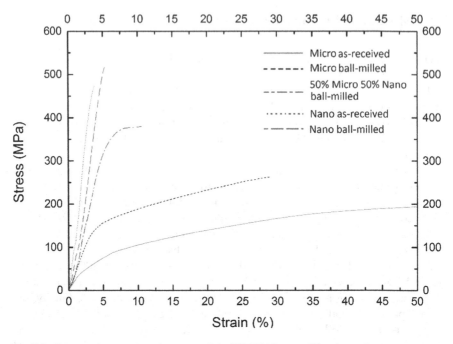

Fig. 7.5 Compressive stress-strain curves of the BP-ECAP consolidated samples

In Fig. 7.5, the results of the compression tests are displayed. The samples originated from the nano-sized powders showed a brittle mechanical behavior but an extremely high strength. As expected, those originated from the micro-sized powders showed a ductile behavior and much lower strength. Again, the mixed sample displayed a mechanical behavior that is a compromise in terms of ductility and strength between the other two ball-milled samples.

7.4 Microstructural Characterization

A typical TEM microstructure of the sample consolidated from the as-received micro Al powder is depicted in Fig. 7.6a. Quasi-equiaxed grains with homogeneous sizes (about 3 μm) are observed in the whole investigated area. Discrete NPs and small clusters are found dispersed within the grains and at the grain boundaries, as shown in Fig. 7.6a and b. EDS analysis revealed that the NPs are oxygen-rich compounds (Fig. 7.6g and h). The NPs were identified as γ-Al_2O_3 by the SAED pattern (Fig. 7.6f). The original oxygen-rich surfaces of the starting particles were not noticeable although they were observed in the samples consolidated at lower temperatures as shown in some previous works [3–5]. Therefore, the observed NPs are expected to be oxide compounds originating from the amorphous oxide layer

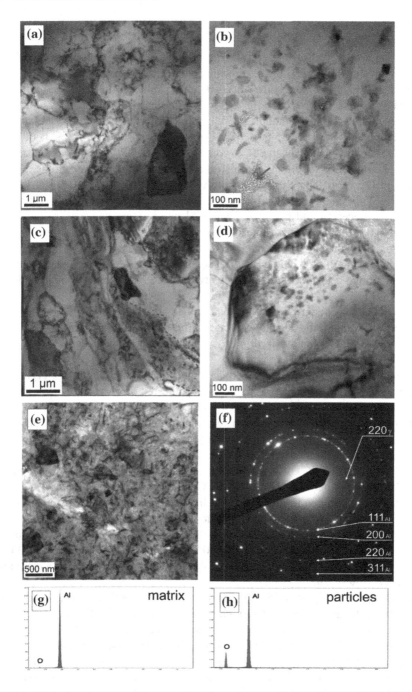

Fig. 7.6 TEM microstructures of the consolidated samples: **a** and **b** from the as-received micro Al powder; **c–e** from the ball milled micro Al powder. **f** SAED pattern identifying the γ-Al$_2$O$_3$ NPs and **g** and **h** EDS spectra of the Al matrix and the NPs, respectively, in the sample consolidated from the as-received micro Al powder [38]

[6–13] that covers the Al particles. This aspect will be further investigated and discussed.

The microstructures of the consolidated sample from the ball-milled micro Al powder are shown in Fig. 7.6c–e, showing coarse grains with average size between 1 and 3 μm (Fig. 7.6c) and some finer structures (indicated in Fig. 7.6c by the red dashed lines). The latter is characterized by a high-density of oxide NPs, submicron Al grains and a high amount of microstructural defects (Fig. 7.6e). Some micro-sized grains, especially those between two zones with high contents of oxide particles, show elongated shape. It is worth noting that NPs are present not only in the areas with finer structures, but also dispersed in some coarser grains, as displayed in Fig. 7.6d. The NPs are again identified as γ-Al_2O_3.

In Fig. 7.7a and b, the microstructures of the consolidated samples produced from the as-received and ball-milled nano Al powders are depicted, respectively. The two microstructures look quite similar and the effect of the ball milling was not as marked as in the case of the micro-sized Al powder. In some areas, small grains (lower than 1 μm in size) are detected. In some other areas, grain boundaries are not easily identified because of the high density of oxide NPs which are again indexed as γ-Al_2O_3 (Fig. 7.7d–f).

The billet consolidated from the ball milled mixture of the micro and nano powders showed a bimodal microstructure (Fig. 7.7c). The micro-sized grains are separated by zones of nano-sized grains and high amount of oxide NPs (indicated by the red dashed lines in Fig. 7.7c).

Finally, it is worth noting that in all the materials, a small number of oxygen-rich amorphous NPs are detected, as indicated by the red arrows in Fig. 7.6a and b. Several reports show that the Al particles are covered by an amorphous oxide layer 2–4 nm thick [3, 6–13], therefore a phase transformation from an amorphous- to a γ-phase occurred during the consolidation process. TG, DSC and more XRD were performed to shed further light on the formation of γ-Al_2O_3 NPs.

In Fig. 7.8a, the DSC curve for the as-received nano Al powder shows an intense endothermic peak between 275 and 370 °C and a weak exothermic peak between 470 and 530 °C. In contrast, in Fig. 7.8b, the DSC curve for the as-received micro Al powder shows just a weak endothermic peak between 391 and 399 °C.

The TG measurement performed on the nano Al powder showed a reduction of mass in the first stage of heating (100–300 °C), a drastic loss of mass at ~ 300 °C corresponding to the DSC endothermic peak and then a region featuring a mass loss at a much slower rate (Fig. 7.9).

XRD was carried out at 25, 200 (below the DSC endothermic peak), 400 (between the DSC endothermic and exothermic peaks) and 600 °C (the ECAP temperature which is above the DSC exothermic peak) to acquire information about phase stability. As mentioned, 600 °C is the ECAP processing temperature.

Figure 7.10a and b display the patterns for the as-received nano Al powder at 25 and 200 °C. The peaks of Al and aluminum trihydroxide γ-Al(OH)$_3$ were identified. The γ-Al(OH)$_3$ phase is also known as gibbsite and possesses a monoclinic symmetry with the space group P2$_1$/n. At 400 °C, the peaks of the hydroxide disappeared and only those of Al were detected (Fig. 7.10c). Finally, at 600 °C the loose

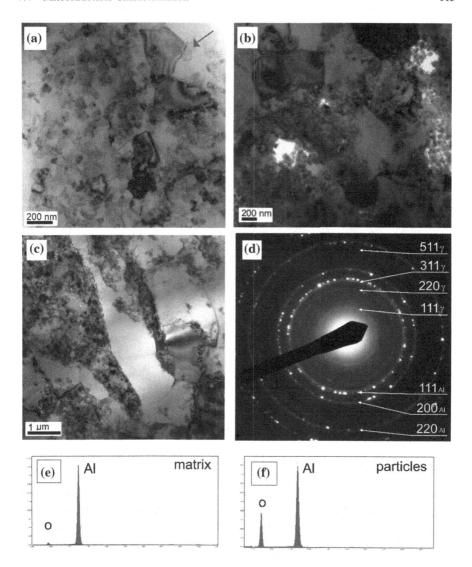

Fig. 7.7 Microstructures of the consolidated samples: **a** from the as-received nano Al powder, **b** from the ball milled nano Al powder, and **c** from the ball-milled mixture of micro and nano Al powders. **d** SAED pattern identifying γ-Al₂O₃ and **e** and **f** EDS spectra of the Al matrix and the NPs [38]

nano Al powder showed both the peaks of Al and γ-Al$_2$O$_3$ phases (Fig. 7.10d). By zooming the XRD pattern at 400 °C, an hump related to an amorphous phase is visible at low incident angles, where the peaks of the Al(OH)$_3$ and γ-Al$_2$O$_3$ are present at 25 and 600 °C, respectively (see for comparison Fig. 7.10e and f). XRD was also performed on the micro Al powder, but the only Al peaks were detected

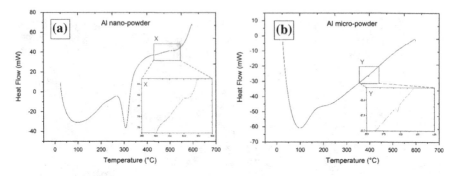

Fig. 7.8 a DSC curve for the nano Al powder with the inset showing a magnified section in area
X. **b** DSC curve for the micro Al powder with the *inset* showing a magnified section in area Y [38]

Fig. 7.9 TG curve for the
nano Al powder [38]

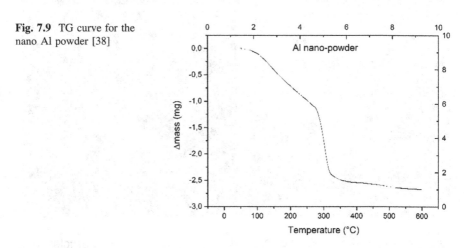

(Fig. 7.11). A shift of the angular peak positions toward lower angles was
noticeable with increasing temperature, most probably due to a thermal drift. The
specific surface of the micro Al particles is much lower than that of the nano Al
particles; therefore, the amount of oxide/hydroxide growing on their surface is
believed to be not enough to be detected.

7.5 Discussion of the Results

Aluminum is thermodynamically unstable with respect to its oxide and hydroxide in
air. Both bulk Al and particles are naturally covered by an amorphous oxide layer
which is about 2–4 nm thick at room temperature [3, 6–13]. This layer is very
compact and stable and protects the interior from further oxidation. Growth of
the amorphous oxide layer is limited by the outward diffusion of Al cations [14].

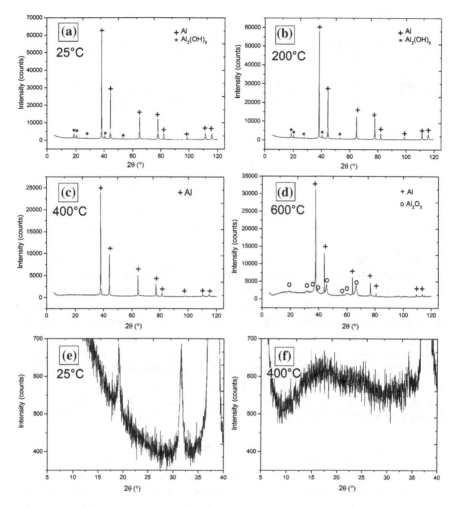

Fig. 7.10 XRD patterns for the as-received nano Al powder: **a** at 25 °C, **b** at 200 °C, **c** at 400 °C, and **d** at 600 °C. Closeups between 5° and 40° are shown **e** at 25 °C and **f** at 400 °C [38]

The amorphous alumina layer becomes metastable when the thickness exceeds a threshold value [10, 15]. Besides, it is largely accepted that the chemical reactivity of aluminum and other metals at the nanometer scale differs from that measurable at the macroscopic level; indeed different transformations and reactions may happen at the nano-scale, which are not favored at the micro-scale [16–20]. The small size and large surface area provide a significantly high chemical reactivity. For example, passive bulk aluminum is practically stable, even in boiling water [9], but nano-sized Al particles react with water to form aluminum hydroxide, releasing H_2 [21]. There is also some evidence that Al trihydroxide can form during prolonged exposure to humid environment in the outermost passive layer of bulk Al samples

Fig. 7.11 XRD pattern for
the as-received micro Al
powder at 25, 200, 400 and
600 °C. A closeup in
correspondence of the [111]
peak is depicted in the *inset*
[38]

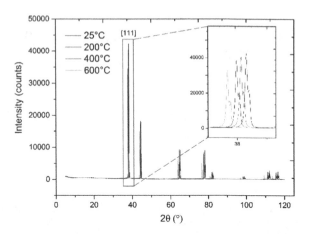

[22–24]. This reaction is likely to happen more decisively on the Al nanoparticles. Moreover, the oxide on the Al nanoparticles is revealed to be more porous than that on the micro-sized Al particles and potentially able to absorb larger amounts of atmospheric moisture [1]. In agreement with these observations, XRD revealed that an aluminum hydroxide layer (gibbsite) covers the Al nanoparticles used in the present experiments.

When the Al NPs are heated, they undergo an endothermic transformation between 275 and 370 °C. At 200 °C, the peaks of gibbsite were still very much visible, whereas at 400 °C they disappeared. The endothermic transformation is attributed to the dehydration of the trihydroxide, causing its transformation into alumina and water according to:

$$2Al(OH)_3 \rightarrow Al_2O_3 + 3H_2O \tag{7.1}$$

The DSC results are in good agreement with some previous works on dehydration of gibbsite [25–28]. The occurrence of dehydration is also supported by the TG results (Fig. 7.9), which show a drastic reduction of weight corresponding to the endothermic peak, confirming the loss of H_2O molecules during the dehydration reaction. Furthermore, the loss in weight visible in the first part of the curve can be ascribed to the evaporation of the adsorbed water. XRD carried out at 400 °C on nano Al particles exhibited only the peaks of the face-centered cubic Al structure. No characteristic peaks for crystalline—Al_2O_3 were noticeable, and a broad hump ascribable to an amorphous phase was visible at low incident angles (see for comparison Fig. 7.10e and f).

Dehydration is a very complex process, which leads to the loss of the long-range order in the crystal structure [29]. The transformations gibbsite $\rightarrow$ amorphous Al_2O_3 and gibbsite $\rightarrow$ boehmite(γ-AlOOH) $\rightarrow$ amorphous Al_2O_3 were proposed for gibbsite particles [9, 27–31]. The formation of the amorphous Al_2O_3 is an intermediate stage between the solid-state reaction from gibbsite to the crystalline alumina. The gibbsite structure is distorted during its dehydration/amorphization, causing the variation and loss of the cell parameters and the loss of the long-range crystal order due to structural reordering and the loss of water [28, 29]. At 600 °C, the peaks of γ-Al_2O_3 were detected in the powder and consolidated samples. Therefore, between 400 and 600 °C the amorphous phase transforms into a crystalline structure. As shown in Fig. 7.8a, a small exothermic peak between 470 and 530 °C was detected by DSC analysis on the nano-Al particles. This peak might be assigned to the amorphous $Al_2O_3 \rightarrow \gamma$-$Al_2O_3$ transformation. However, it is worth noting that some other phenomena could cause exothermic heat release, namely residual oxidation, annealing of defects or strain relaxation in the Al particle core. Accelerated oxidation was observed by Trunov et al. [6] in a narrow range from 550 to 650 °C below the Al melting temperature, when a critical value of the oxide thickness (4 nm) is exceeded. It is suggested that the newly formed γ-Al_2O_3 has a density 20 % higher than that of the amorphous—Al_2O_3. Thus, the cubic alumina does not fully cover the Al particle surface. The formation of bare Al spots and their immediate oxidation may cause the rapid increase of oxide until a continuous layer is formed. It is estimated that the thickness of the γ-Al_2O_3 layer is about 15–20 nm. It is noted that on the TG curve in Fig. 7.9, the increase in mass is not evident because the test was carried out in Ar atmosphere. Thus, at 600 °C, the content of Al oxide is expected to be higher than that in the as-received powder and the amount of oxide in the consolidated sample has to be even higher.

The diffraction peaks of Al_2O_3 and $Al(OH)_3$ were detected only for the nano Al powder. The amount of these phases is believed to be too low to detect in the micro Al powder. A weak endothermic DSC peak was noticeable between 391 and 399 °C when the micro-sized powder underwent the heating scan (Fig. 7.8b). This might suggest the presence of some hydroxide on the surface that transformed into amorphous Al_2O_3, but there is no XRD evidence. Nonetheless, according to the results achieved using the nano Al particles, no matter whether the Al particles were covered by the hydroxide at low temperatures (<400 °C), at the consolidation temperature of 600 °C, a γ-Al_2O_3 layer is supposed to cover the metal particles.

In Figs. 7.6 and 7.7, it was shown that the oxide phase was not present as boundaries between the original Al particles, as previously found in samples consolidated at 400 °C [3] or even lower temperatures [4, 5]. The oxide was found, indeed, in the form of particles well dispersed in the metal matrix in all the samples investigated. γ-Al_2O_3 NPs were also obtained by Balog et al. [3] after annealing of the consolidated micro Al powder (1–9 μm) at temperatures higher than 450 °C. The driving force for the transformation from oxide layers into oxide particles is considered to be the minimization of surface energy, which is made possible by diffusion at the higher temperatures. The ECAP process also play a role in dispersing the NPs [32]. The transformation process from the oxide layer into discrete γ-Al_2O_3 NPs can be summarize in eight steps, as illustrated in Fig. 7.12.

The microstructures observed in the present study correspond well with the mechanical properties obtained. The very high content of reinforcing γ-Al_2O_3 NPs in the materials produced from nano Al particles plausibly leads to their extraordinary high hardness (about 166 HV), low ductility and high compressive strength (Figs. 7.4 and 7.5). For comparison, it is worth reporting that the Vickers hardness

Phase transformations during heating

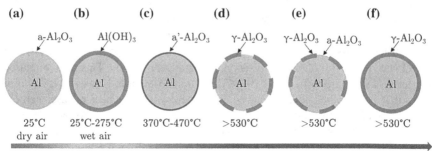

ECAP consolidation at 600°C

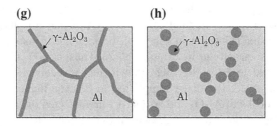

Fig. 7.12 Schematic of the transformation of the oxide layer: **a** A nano Al particle is covered by an amorphous oxide layer 2–4 nm thick at room temperature (a-Al_2O_3), **b** as soon as they are exposed to atmospheric moisture the oxide reacts with water to form aluminum hydroxide $Al(OH)_3$, **c** between 275 and 370 °C, the hydroxide dehydrates and loose its long range crystallographic order, forming amorphous alumina (a'-Al_2O_3) which remains stable below 470 °C, **d** between 470 and 530 °C, a'-Al_2O_3 transforms into γ-Al_2O_3 which has a density 20 % higher than that of a-Al_2O_3, leading to its not fully covering the Al particle surface, **e–f** the bare Al spots and their immediate oxidation may cause the rapid increase of oxide until a continuous layer is formed, and **g–h** when the nano Al particles are consolidated by ECAP at 600 °C, the continuous γ-Al_2O_3 layers transform into discrete γ-Al_2O_3 NPs to minimize the surface energy. The ECAP process plays a role in dispersing the NPs [38]

of coarse-grained and UFG pure aluminum are about 20 and 40 HV, respectively [2]. The in situ reinforcement led to a hardness increase of about 400 % when compared with the unreinforced UFG Al.

The samples produced from the micro Al powder, reinforced with lower amounts of alumina NPs, showed much lower hardness and strength but higher ductility. The samples prepared from the as-received powder showed a hardness of about 29 HV and a yield strength of 40 MPa while those prepared using the ball-milled powder exhibited a hardness of 62 HV and a yield stress of 140 MPa. The hardness values are consistent with the results achieved by consolidating Al particles at lower temperature. The as-received and ball-milled Al powder consolidated by ECAP for 4 passes at 400 °C showed higher Vickers hardness (37 and

93 HV, respectively) due to a softer annealing action exerted during holding at the processing temperature. Finally, the sample produced from the mixture of nano- and micro-particles showed a hardness of about 120 HV, a yield stress of about 350 MPa and a ductility of 11 %, representing a compromise between those achieved by the samples produced using the ball-milled micro-sized Al powder and the ball-milled nano-sized Al powder. This might suggest the possibility of using different proportions of nano- and micro-particles in order to produce the desired combinations of mechanical properties.

The enhanced hardness and strength are due to the strengthening effect of the dispersed nanoparticles and of the refined microstructure. The nanoparticles can interact with dislocations, hampering their movement and leading to dislocations bowing around the particles (Orowan strengthening) [33, 34]. The mismatch in the coefficient of thermal expansion between the reinforcement and the metal matrix is also believed to contribute to strengthening since it leads to the formation of dislocations, which are geometrically necessary to accommodate the different contractions [35]. Moreover, the grain size has a strong influence on strength since the grain boundaries can obstruct dislocation movement [36, 37], and the particles play an important role in refining grains in the composites because they can hinder grain growth at high temperatures by acting as pinning points for grain boundary migration. The effect of the finer grain size is demonstrated in the sample produced from ball milled micro Al powder, which showed higher strength and hardness than that produced from as-received powder.

References

1. A.L. Ramaswamy, P. Kaste, S.F. Trevino, A "Micro-vision" of the physio-chemical phenomena occurring in nanoparticles of aluminum. J. Energ. Mater. 22(1), 1–24 (2004). http://www.tandfonline.com/doi/abs/10.1080/07370650490438266
2. M. Saravanan, R.M. Pillai, B.C. Pai, M. Brahmakumar, K.R. Ravi, Equal channel angular pressing of pure aluminium—an analysis. Bull. Mater. Sci. 29, 679–684 (2006)
3. M. Balog, P. Krizik, M. Nosko, Z. Hajovska, M.V. Castro Riglos, W. Rajner, D.S. Liu, F. Simancik, Forged HITEMAL: Al-based MMCs strengthen with nanometric thick Al_2O_3 skeleton. Mater. Sci. Eng. A. 613, 82–90 (2014)
4. X. Wu, W. Xu, K. Xia, Pure aluminum with different grain size distributions by consolidation of particles using equal-channel angular pressing with back pressure. Mater. Sci. Eng. A 493, 241–245 (2008)
5. W. Xu, X. Wu, T. Honma, S.P. Ringer, K. Xia, Nanostructured $Al-Al_2O_3$ composite formed in situ during consolidation of ultrafine Al particles by back pressure equal channel angular pressing. Acta Mater. 57, 4321–4330 (2009)
6. M.A. Trunov, M. Schoenitz, X. Zhu, E.L. Dreizin, Effect of polymorphic phase transformations in Al_2O_3 film on oxidation kinetics of aluminum powders. Combust. Flame 140, 310–318 (2005)
7. X. Phung, J. Groza, E.A. Stach, L.N. Williams, S.B. Ritchey, Surface characterization of metal nanoparticles. Mater. Sci. Eng. A 359(1–2), 261–268 (2003). http://www.sciencedirect.com/science/article/pii/S0921509303003484

8. M. Balog, F. Simancik, M. Walcher, W. Rajner, C. Poletti, Extruded Al–Al$_2$O$_3$ composites formed in situ during consolidation of ultrafine Al powders: effect of the powder surface area. Mater. Sci. Eng. A **529**, 131–137 (2011)

9. K. Wafers, C. Misra, Oxides and hydroxides of aluminum. Alcoa Technical Report No. 19 Revised, Alcoa Laboratories, 64 (1987)

10. B. Rufino, F. Boulc'h, M.V. Coulet, G. Lacroix, R. Denoyel, Influence of particles size on thermal properties of aluminium powder. Acta Materialia. **55**, 2815–2827 (2007)

11. J.C. Sanchez-Lopez, A.R. Gonzalez-Elipe, A. Fernandez, Passivation of nanocrystalline Al prepared by the gas phase condensation method: an X-ray photoelectron spectroscopy study. J. Mater. Res. **13**, 703–710 (1998)

12. P.E. Doherty, R.S. Davis, Direct observation of the oxidation of aluminum single-crystal surfaces. J. Appl. Phys. **34**, 619–628 (1963)

13. K. Tomas, M.W. Roberts, Direct observation in the electron microscope of oxide layers on aluminum. J. Appl. Phys. **32**, 70–75 (1961)

14. L.P.H. Jeurgens, W.G. Sloof, F.D. Tichelaar, E.J. Mittemeijer, Growth kinetics and mechanisms of aluminum-oxide films formed by thermal oxidation of aluminum. J. Appl. Phys. **92**, 1649–1656 (2002)

15. L.P.H. Jeurgens, W.G. Sloof, F.D. Tichelaar, E.J. Mittemeijer, Thermodynamic stability of amorphous oxide films on metals: application to aluminum oxide films on aluminum substrates. Phys. Rev. B. **62**, 4707–4719 (2000)

16. L. Meda, G. Marra, L. Galfetti, F. Severini, L. De Luca, Nano-aluminum as energetic material for rocket propellants. Mater. Sci. Eng. C **27**, 1393–1396 (2007)

17. B.J. Henz, T. Hawa, M.R. Zachariah, On the role of built-in electric fields on the ignition of oxide coated nanoaluminum ion mobility versus Fickian diffusion. J. Appl. Phys. **107**, 024901 (2010)

18. M. Valden, X. Lai, D.W. Goodman, Onset of catalytic activity of gold clusters on titania with the appearance of nonmetallic properties. Science **281**, 1647–1650 (1998)

19. Y. Yang, S. Wang, Z. Sun, D.D. Dlott, Near-infrared laser ablation of poly tetrafluoroethylene (Teflon) sensitized by nanoenergetic materials. Appl. Phys. Lett. **85**, 1493–1495 (2004)

20. B. Yoon, H. Hakkinen, U. Landman, A.S. Worz, J.M. Antonietti, S. Abbet, K. Judai, U. Heiz, Charging effects on bonding and catalyzed oxidation of CO on Au$_8$ clusters on MgO. Science **307**, 403–407 (2005)

21. B. Alinejad, K. Mahmoodi, Hydrogen generation from water and aluminum promoted by sodium stannate. Int. J. Hydrogen Energy **34**, 7934–7938 (2009)

22. B. Strohmeier, An ESCA method for determining the oxide thickness on aluminum alloys. Surf. Interface Anal. **15**, 51–56 (1990)

23. N.A. Thorne, P. Thuery, A. Frichet, P. Gimenez, A. Sartre, Hydration of oxide films on aluminium and its relation to polymer adhesion. Surf. Interface Anal. **18**, 236240 (1990)

24. M. Amstutz, M. Textor, Applications of surface-analytical techniques to aluminium surfaces in commercial semifabricated and finished products. Surf. Interface Anal. **19**, 595–600 (1992)

25. D.T.Y. Chen, DSC dehydration peaks and solubility products of Al(OH)$_3$. Thermochim. Acta **11**, 101–104 (1975)

26. J.M.R. Mercury, P. Pena, A.H. De Aza, D. Sheptyakov, X. Turrillas, On the decomposition of synthetic gibbsite studied by neutron thermodiffractometry. J. Am. Ceramic Soc. **89**, 3728–3733 (2006)

27. A.D.V. Souza, C.C. Arruda, L. Fernandes, M.L.P. Antunes, P.K.K. Kiyohara, R. Salomao, Characterization of aluminum hydroxide (Al(OH)$_3$) for use as a porogenic agent in castable ceramics. J. Eur. Ceramic Soc. **35**, 803–812 (2015)

28. B.K. Gan, I.C. Madsen, J.G. Hockridge, In situ X-ray diffraction of the transformation of gibbsite to alfa-alumina through calcination: effect of particle size and heating rate. J. Appl. Crystallogr. **42**, 697–705 (2009)

29. H. Wang, B. Xu, P. Smith, M. Davies, L. De Silva, C. Wingate, Kinetic modelling of gibbsite dehydration/amorphization in the temperature range 823–923 K. J. Phys. Chem. Solids **67**, 2567–2582 (2006)

30. B. Whittington, D. Ilievski, Determination of the gibbsite dehydration reaction pathway at conditions relevant to Bayer refineries. Chem. Eng. J. **98**, 89 (2004)
31. J. Rouquerol, F. Rouquerol, M. Granteaume, Thermal decomposition of gibbsite under low pressures: I. Formation of the boehmitic phase. J. Catal. **36**, 99–110 (1975)
32. R. Lapovok, D. Tomus, C. Bettles, Shear deformation with imposed hydrostatic pressure for enhanced compaction of powder. Scripta Mater. **58**, 898–901 (2008)
33. D. Hull, D.J. Bacon, *Introduction to Dislocations*, 4th edn. (Butterworth-Heinemann, 2001)
34. Z. Zhang, D.L. Chen, Contribution of Orowan strengthening effect in particulate-reinforced metal matrix nanocomposites. Mater. Sci. Eng. A **483–484**, 148–152 (2008)
35. R.J. Arsenault, N. Shi, Dislocation generation due to differences between the coefficients of thermal-expansion. Mater. Sci. Eng. A **81**, 175–187 (1986)
36. E.O. Hall, The deformation and aging of mild steel. Proc. Phys. Soc. London, Sect. B **64**, 747–753 (1951)
37. N.J. Petch, The cleavage strength of polycrystals. J. Iron Steel Res. **174**, 25–28 (1953)
38. R. Casati, A. Fabrizi, A. Tuissi, K. Xia, M. Vedani, ECAP consolidation of Al matrix composites reinforced with in-situ γ-Al_2O_3 nanoparticles. Mater. Sci. Eng. A**648**, 113–122 (2015)

Chapter 8
Conclusions

Different powder metallurgy routes were designed and investigated to produce Al matrix composites reinforced with Al_2O_3 nanoparticles. They relied on high-energy ball milling and consolidation via hot extrusion and ECAP.

Ball milling powder processing was used to break the alumina nanoparticles clusters. It was also aimed at dispersing the ceramic reinforcement throughout the metal matrix discretely. Moreover, it turned out to be able to lead to the in situ production of nanoparticles by fragmentation of the passivation oxide layer that covers the aluminum powder particles and to the formation of very thin Al_2O_3 dispersoids.

ECAP and hot extrusion revealed to be optimal powder consolidation methods, able to produce fully dense materials even after a single pass at temperature lower than those used in conventional sintering. The consolidation occurred almost instantaneously as the powder passed through the die. Therefore, low temperature, short processing time and NPs dispersed in the matrix, which may act as pinning points for grain growth, led to a very fine microstructure. The MMnCs billets prepared by extrusions were also cold rolled down to small section wires (1 mm × 1 mm). They showed a fairly good formability since they were cold worked without any evidence of cracks.

The MMnCs exhibited extremely high mechanical properties. For example, the strength of the in situ reinforced nanocomposite produced by ECAP was 302 MPa, while an even higher value was achieved by the in situ/ex situ reinforced nanocomposite, corresponding to 385 MPa. The material reinforced with 2 wt% of Al_2O_3 consolidated via hot extrusion and then cold rolled reached YS and UTS values of 282 and 373 MPa, respectively, while the wire produced from ball-milled CP Al powder reached YS and UTS of 225 and 302 MPa, respectively. The Al nanocomposites also showed extremely high hardness values ranging from 96 to 199 HV. This latter value was reached by the Al sample reinforced with 20 % of Al_2O_3 NPs.

© The Author(s) 2016
R. Casati, *Aluminum Matrix Composites Reinforced with Alumina Nanoparticles*,
PoliMI SpringerBriefs, DOI 10.1007/978-3-319-27732-5_8

Internal friction in the nanocomposites wires was determined at 0.1, 1 and 10 Hz. The tanδ curves show well defined peaks. Higher amount of Al_2O_3 NPs led to higher damping capacity. The relaxation peak shifts towards higher temperatures with increasing loading frequency. The storage modulus of the MMnC reinforced with ex situ NPs is higher than that of the MMnC reinforced with only in situ NPs under the testing conditions.

Al nanoparticles, replacing the Al micro-particles, were also investigated in a second stage. The Al nanoparticles have been consolidated at 600 °C by using ECAP with back pressure. The Al NPs revealed to be covered by Al hydroxide, which first transformed into amourphous-Al_2O_3, then into γ-Al_2O_3. The consolidated samples showed exceptionally high hardness (166 HV), even without any addition of ex situ alumina. However, the nanocomposites produced using Al NPs revealed brittle compressive behavior. Al NPs were then mixed with 50 % of micro-sized Al to increase the material ductility in search of a more suitable compromise of properties.

Some suggestions for potential future works are hereunder introduced.

- Nanocomposites with different matrix and particles might be produced using the same PM routes presented in this work. In particular, it would be interesting preparing commercial Al-alloy wires or sheets reinforced with nanoparticles, as examples of industrial applications.
- The study of the creep behavior and the structural characterization at high temperature of MMnCs should be more intensely studied. MMnCs have been indeed proposed as suitable materials for high temperature applications, thanks to the thermodynamic stability, though there is a lack of knowledge about this topic.
- MMnCs reinforced with NPs prepared in situ revealed a good compromise of properties in terms of ductility and strength. Different amount of reinforcement might be formed by controlled oxidation of the Al powder at different temperatures and for different times.
- Precursors with high amount of NPs can be prepared by using ball milling and cold or hot pressing, so as to be employed in a semi-solid casting process or directly in liquid casting processes.
- Al matrix composites produced using Al NPs or using Al micro-particles reinforced with high amount of ceramic NPs showed very high hardness and low ductility, making difficult their application. However, they could be deposited on a metal surface, for examples by means of laser cladding, in order to improve the wear properties of the bulk metal.
- The use of a mixture of Al NPs and Al micro-particles could allow tuning the amount of reinforcement and the corresponding expected properties, by adding a further degree of freedom in the synthesis process of MMnCs.

Printed in the United States
By Bookmasters